International E-Conference on Computer Science 2005

LECTURE SERIES
ON COMPUTER AND COMPUTATIONAL SCIENCES II

International E-Conference on Computer Science 2005

Edited by T. E. Simos and G. Psihoyios

LONDON AND NEW YORK

First published 2005 by VSP

2 Park Square, Milton Park, Abingdon, Oxfordshire OX14 4RN
52 Vanderbilt Avenue, New York, NY 10017

Routledge is an imprint of the Taylor & Francis Group, an informa business

First issued in hardback 2019

A C.I.P. record for this book is available from the Library of Congress

ISBN 13: 978-90-6764-425-9 (pbk)
ISBN 13: 978-1-138-41299-6 (hbk)

Brill Academic Publishers
P.O. Box 9000, 2300 PA Leiden,
The Netherlands

*Lecture Series on Computer
and Computational Sciences*
Volume 2, 2005, pp. i-ii

Preface for the Proceedings of the

International e-Conference on Computer Science 2005 (IeCCS 2005)

Recognised Conference by the European Society of Computational Methods in Sciences and Engineering (ESCMSE)

This volume of the *Lecture Series on Computer and Computational Sciences* contains the refereed *Proceedings of the International e-Conference on Computer Science (IeCCS-2005)*, which took place in the internet between 19 and 31 May 2005.

The international electronic conference IeCCS-2005 is the first in a new series of e-conferences and it proved a truly successful event that attracted more than 90 papers from all over the world. The nearly 40 short papers or extended abstracts contained in this volume represent those short papers that were accepted for publication after an appropriate refereeing process, in accordance with established practices.

We are pleased with the overall standard of the included extended abstracts and we believe that readers will enjoy the contributions contained herein. The conference contributors have also submitted full-length paper versions of their extended abstracts and an appropriate selection of these full-length papers will be published in the journal *Computing Letters*, after a further refereeing process.

We would like to thank all participants for their contributions and the members of the scientific committee for their support. We would also like to warmly thank the plenary speakers who agreed to undertake such a laborious function in a rather short notice.

We would like also to thank:

- The Scientific Committee of IeCCS 2005 (see in page iv for the Conference Details) for their help and their important support.
- The Symposiums' Organisers for their excellent editorial work and their efforts for the success of IeCCS 2005.
- The invited speakers for their acceptance to give keynote lectures.

- The Organising Committee for their help and activities for the success of IeCCS 2005.

- Special thanks for the Secretary of IeCCS 2005, Mrs Eleni Ralli-Simou (which is also the Administrative Secretary of the European Society of Computational Methods in Sciences and Engineering (ESCMSE)) for her excellent job.

Prof. Theodore Simos
President of ESCMSE
Chairman IeCCS 2005
Editor of the Proceedings
Department of Computer Science
and Technology
University of the Peloponnese
Tripolis
Greece
E-mail: tsimos@mail.ariadne-t.gr

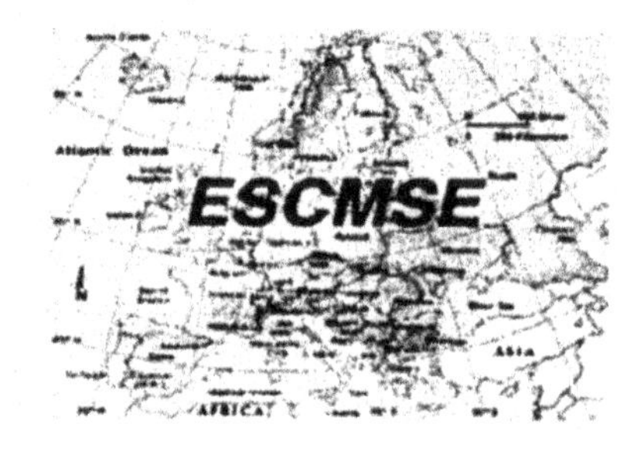

G. Psihoyios[b]
Co-Chairman IeCCS 2005
Co-Editor of the Proceedings
[b] APU, Cambridge, UK
[b] Correspondence address: 192 Campkin Road, Cambridge CB4 2LH, UK
E-mail: g.psihoyios@ntlworld.com

June 2005

Brill Academic Publishers
P.O. Box 9000, 2300 PA Leiden.
The Netherlands

Lecture Series on Computer and Computational Sciences
Volume 2, 2005, pp. iii-iv

Conference Details

International e-Conference on Computer Science (IeCCS 2005)

Recognised Conference by the European Society of Computational Methods in Sciences and Engineering (ESCMSE)

Chairmen and Organisers

Professor T.E. Simos, President of the European Society of Computational. Methods in Sciences and Engineering (ESCMSE). Active Member of the European Academy of Sciences and Arts and Corresponding Member of the European Academy of Sciences, Department of Computer Science and Technology, Faculty of Sciences and Technology, University of Peloponnese, Greece.

Dr. Georgios Psihoyios, Vice-President of the European Society of Computational. Methods in Sciences and Engineering (ESCMSE). FIMA. APU, Cambridge, UK

Scientific Committee

Prof. Dr. Gerik Scheuermann, Germany
Prof. James Hendler, USA
Prof. Sanguthevar Rajasekaran, USA
Prof. Reda A. Ammar, USA
Prof. Vijay K. Vaishnavi, USA
Prof. Vijay Kumar, USA
Prof. Colette Rolland, France
Prof. Michel Daydé, France
Prof. Pascal Sainrat, France
Dr. Peter Lucas, The Netherlands
Prof. Richard Squier, USA
Prof. dr. H.A. (Erik) Proper, The Netherlands
Dr. Len Freeman, UK

Invited Speakers

Prof. Reda A. Ammar, USA
Prof. dr. H.A. (Erik) Proper, The Netherlands

Organizing Committee

Mrs Eleni Ralli-Simou (Secretary of IeCCS 2005)
Mr. D. Sakas
Mr. Z. A. Anastassi
Mr. Th. Monovasilis
Mr. K. Tselios
Mr. G. Vourganas

Brill Academic Publishers
P.O. Box 9000, 2300 PA Leiden,
The Netherlands

Lecture Series on Computer and Computational Sciences
Volume 2, 2005, pp. v-vi

European Society of Computational Methods in Sciences and Engineering (ESCMSE)

Aims and Scope

The ***European Society of Computational Methods in Sciences and Engineering (ESCMSE)*** is a non-profit organization. The URL address is: **http://www.uop.gr/escmse/**

The aims and scopes of ***ESCMSE*** is the construction, development and analysis of computational, numerical and mathematical methods and their application in the sciences and engineering.

In order to achieve this, the ***ESCMSE*** pursues the following activities:

- Research cooperation between scientists in the above subject.
- Foundation, development and organization of national and international conferences, workshops, seminars, schools, symposiums.
- Special issues of scientific journals.
- Dissemination of the research results.
- Participation and possible representation of Greece and the European Union at the events and activities of international scientific organizations on the same or similar subject.
- Collection of reference material relative to the aims and scope of ***ESCMSE.***

Based on the above activities, ***ESCMSE*** has already developed an international scientific journal called **Applied Numerical Analysis and Computational Mathematics (ANACM)**. This is in cooperation with the international leading publisher, **Wiley-VCH**.

ANACM is the official journal of ***ESCMSE.*** As such, each member of ***ESCMSE*** will receive the volumes of **ANACM** free of charge.

Categories of Membership

European Society of Computational Methods in Sciences and Engineering (ESCMSE)

Initially the categories of membership will be:

- **Full Member (MESCMSE):** PhD graduates (or equivalent) in computational or numerical or mathematical methods with applications in sciences and engineering, or others who have contributed to the advancement of computational or numerical or mathematical methods with applications in sciences and engineering through research or education. Full Members may use the title MESCMSE.

• **Associate Member (AMESCMSE):** Educators, or others, such as distinguished amateur scientists, who have demonstrated dedication to the advancement of computational or numerical or mathematical methods with applications in sciences and engineering may be elected as Associate Members. Associate Members may use the title AMESCMSE.

• **Student Member (SMESCMSE):** Undergraduate or graduate students working towards a degree in computational or numerical or mathematical methods with applications in sciences and engineering or a related subject may be elected as Student Members as long as they remain students. The Student Members may use the title SMESCMSE

• **Corporate Member:** Any registered company, institution, association or other organization may apply to become a Corporate Member of the Society.

Remarks:

1. After three years of full membership of the European Society of Computational Methods in Sciences and Engineering, members can request promotion to Fellow of the European Society of Computational Methods in Sciences and Engineering. The election is based on international peer-review. After the election of the initial Fellows of the European Society of Computational Methods in Sciences and Engineering, another requirement for the election to the Category of Fellow will be the nomination of the applicant by at least two (2) Fellows of the European Society of Computational Methods in Sciences and Engineering.

2. All grades of members other than Students are entitled to vote in Society ballots.

3. All grades of membership other than Student Members receive the official journal of the ESCMSE Applied Numerical Analysis and Computational Mathematics (ANACM) as part of their membership. Student Members may purchase a subscription to ANACM at a reduced rate.

We invite you to become part of this exciting new international project and participate in the promotion and exchange of ideas in your field.

Brill Academic Publishers
P.O. Box 9000, 2300 PA Leiden,
The Netherlands

Lecture Series on Computer and Computational Sciences
Volume 2, 2005, pp. vii-viii

Table of Contents

Computers & Security Special Session

Brill Academic Publishers
P.O. Box 9000, 2300 PA Leiden,
The Netherlands

Lecture Series on Computer and Computational Sciences
Volume 2, 2005, pp. 1-5

Heuristic Scheduling Algorithms for Stochastic Tasks in a Distributed Multiprocessor Environments

Ehab Abdel Maksoud* and Reda A. Ammar**

* Faculty of Engineering, Cairo University, Egypt
** Computer Science and Engineering Department, University of Connecticut

Received 28 May, 2005; accepted in revised form 2 June, 2005

Abstract: A task-to-processor mapping that balances the workload of the processors will typically increase the overall system throughput and will reduce the computations execution times. In our previous work, we considered the scheduling problem for stochastic tasks, where the time cost distributions of these tasks are known a priori and we developed a heuristic scheduling algorithm that balances the processors loads, based on the load averaging. We developed a new scheduling algorithm using task partitioning whenever needed to improve the results produced by the previous approach. Both algorithms are presented in this paper.

Keywords: real-time systems, multi-processor systems, stochastic tasks, scheduling algorithms, load balancing

1. Introduction

Since the scheduling problem is NP-hard [2], heuristic approaches [1] are used to generate reasonable solutions in an affordable amount of computation time. This can be done either statically or dynamically, depending on when the task assignment decisions are made. Most of published heuristic algorithms [2-9] assumed that each task requires equal execution time and the objective for each of them is to schedule tasks so that each processor has the same number of tasks to maintain a balanced load among the available processors. This contradicts with the nature of most real-time applications which consists of tasks with different execution times. Moreover, the task itself may have different execution times for different inputs. In these applications, the time cost distributions of different tasks, instead of their actual execution times, can be easily estimated [6].

In a previous work [1], we presented a heuristic algorithm, which is based on load balancing, to schedule a set of independent stochastic tasks. In this work, we assumed that the time cost distributions of different tasks are known a priori. We extended the previous approach by allowing task partitioning if it is necessary. The proposed algorithm provides a (nearly) flat distribution, so that all the available processors share, approximately, an equal computational workload. Simulation studies are used to evaluate the merits of the new algorithm in generating a well-balanced load relative to the previous algorithm [1].

2. The modeling approach

We assume a multiprocessor system consist of m identical processors. Tasks arrive from a task source (TS) through a queue (Q) and get fed into a task allocation/scheduling mechanism, or the scheduler, which assigns each task to one of the m processors. The communication between the scheduler and the different processors is through the dispatch queues, $Q_1, Q_2,$, Q_m. Each processor has its own dispatch queue. This organization ensures that each processor will always find some tasks in its dispatch queue when they finish the execution of their current tasks. The scheduler will run in parallel with the processors, scheduling the ready tasks and periodically updating the dispatch queues.

We assume that the number of tasks to schedule is much greater than the number of processors (n > m). We also assume:

1. The time cost distribution for each task T_i, i=1,2,..,n, is known with mean, μ_i, and variance, V_i. μ_i^k and V_i^k are the worst case mean and variance of T_i when it runs on k processors in parallel, where k, the degree of parallelism of the task.
2. When a task is parallelized, all of its parallel partitions have to start simultaneously.

3. Tasks are non-preemptable, i.e., once a task or a partition starts execution, it finishes to its completion.
4. For each task T_i, the worst case mean and variance for any j and k with j<k satisfy the sub-linear speedup assumptions $j * \mu_i^j \le k * \mu_i^k$ and $j * V_i^j \le k * V_i^k$. The sub-linearity is due to the overheads associated with communication and synchronization among the split tasks of a task.

In a previous work [2], we presented a heuristic algorithm to schedule a set of independent stochastic tasks that can not be partitioned, based on load balancing. Using parallelism in tasks can help to avoid the existence of overloaded processors and, thus, obtain better load balancing compared to the previous scheduling algorithm [2]. Therefore, our goal is to develop a simple heuristic scheduling algorithm which exploits parallelism in some tasks, if it is necessary, to yield an appropriate balanced load among the available processors in the system.

3. The Proposed Algorithm

3.1. Terminology

Theorem [18]: If $T_1, T_2, \ldots, T_n$, are n independent random variables with mean values $\mu_1, \mu_2, \ldots, \mu_n$ and variances $V_1, V_2, \ldots, V_n$, then the sum of these variables has the mean $\mu_s = \sum_{i=1}^{n} \mu_i$, the variance $V_s = \sum_{i=1}^{n} V_i$ and the standard deviation $\sigma_s = \sqrt{V_S}$. Based on this theorem:

1. The time cost distribution of the average load for each processor in a system, with m processors, to execute a set of n independent tasks, has the mean μ and the variance V, which are defined as

$$\mu = \frac{\sum_{i=1}^{n} \mu_i}{m} \quad \text{and} \quad V = \frac{\sum_{i=1}^{n} V_i}{m}$$

2. The time cost distribution, T_{P_j}, for the current load of a processor P_j, j=1,2, ...,m, has the mean μ_{P_j} and the variance V_{P_j}, which are defined, respectively, as the sum of the means and the variances for the time cost distributions of tasks or split tasks that are scheduled on that processor, including the current task (or split task).
3. A processor P_j, j=1,2, ...,m, is said to be underloaded if $\mu_{P_j} + \sigma_{P_j} < \mu + 2\sigma$. With equal constraint, the processor takes its full load. Otherwise, the processor is overloaded.
4. FTP_j is the finishing time for the last task (or split task) in the processor P_j, j=1,2, ...,m. The distribution of this time has the mean μ_{FTP_j} and the variance V_{FTP_j} which are defined respectively, as the sum of the means and the variances for the time cost distributions of all tasks or partitions that are scheduled on that processor in the proposed scheduling.
5. FTS is the finishing time of a schedule which is the time required for the last processor to finish executing its assigned task (or split task). The distribution of this time is equal to the finishing time distribution of the processor, say P_k, that has maximum mean. This time cost has the mean μ_{FTS} and the variance V_{FTS} where
6. $\mu_{FTS} = \max_{P_j} \mu_{FTP_j} = \mu_{FTP_k}$ and $V_{FTS} = V_{FTP_k}$.

3.2. Scheduling without task partitioning

1. Evaluate the expected time cost of the average load for each processor in the system, μ and its variance, $\sigma 2$.
2. Order the current tasks (in the task queue (Q)) in a decreasing order of their expected time costs. If more than one have the same expected time cost, order them in a decreasing order of their variances.
3. Set two pointers i and j to indicate the task index in the resulting task queue and the processor index, respectively.
4. For i = 1 to m, allocate a task Ti to a different processor Pj

5. Check for the processor's load: If a processor is under loaded $\mu Pj < \mu + \sigma$, it should take more tasks in the next steps. Exclude processors which are full or overloaded.
6. Add the time cost distribution of the candidate task, Ti, to the current load for each under loaded processor in the system.
7. Evaluate the expected time cost, μ'_P the variance, $\sigma^2_{P_j}$, and the ratio $\left(\frac{\mu - \mu'_{P_j}}{\sigma'_{P_j}}\right)$ for the resultant time cost distribution of each processor, Pj, $j \in \{1,2,\ldots,m\}$.
8. Select the processor that has the maximum value of the ratio and add the target task to it.
9. update the expected time cost and the variance for the current load of the chosen processor. Use step 5 to check the processor's load.
10. Increment i and repeat steps 7 to 10 until i = n.
11. Evaluate the finishing time distribution of the processor with the maximum mean to be the distribution for the finishing time of the schedule, FTS. If more than one distribution have the same maximum, choose the one of maximum variance.

We conducted extensive simulation studies. Since we are interested in whether or not each processor contributes an equal computational workload. Therefore, the most appropriate metric to evaluate the merits of the algorithm is the probability of finishing time of the schedule, FTS to meet the desired schedule time, TS, i.e., P(FTS≤TS). In this study, we assumed that $TS = \mu + 2\sigma$. Since the FTS is given by the sum of several independent random variables, it follows approximately normal distribution with mean μFTS and variance according to the central limit theorem. Therefore, μFTS and were used to find the required probability from the standard normal tables.

No of Processors	$10 \le \mu < 50$ $5 \le \sigma^2 < 10$	$50 \le \mu < 100$ $10 \le \sigma^2 < 20$	$100 \le \mu < 150$ $20 \le \sigma^2 < 30$	$150 \le \mu < 300$ $30 \le \sigma^2 < 60$	$300 \le \mu < 500$ $50 \le \sigma^2 < 100$
4	94.6%	92.3%	88.6%	85.1%	81.7%
8	89.1%	86.2%	83.4%	81.4%	79.9%
16	85.2%	84.1%	80.2%	79.4%	78.5%
24	83.6%	79.5%	78.9%	77.2%	76.7%
32	80.4%	78.7%	77.3%	76.6%	75.3%

3.3. Scheduling with tasks' partitioning

1. Evaluate the expected time cost of the average current load for each processor in the system μ, and its variance V. Order the current tasks (in the task queue (Q)) in an increasing order of their expected time costs. If more than one have the same expected time cost, order them in an increasing order of their variances.
2. Set three pointers i, j, and k to indicate the task index in the resulting task queue, the processor index, and the number of split tasks permitted for the candidate task, respectively. The pointer i changes from 1 to n, j changes from 1 to m, and k changes from 2 to m (or to the max. partition number).
3. For i = 1 to m, schedule a task Ti on a different processor Pj, j = 1,2, ,m.
4. Construct the candidate processor list that gives the available processors in an increasing order of their current loads $(\mu_{P_j} + \sigma_{P_j})$ j=1,2, ,m.
5. Set the pointer i to be m+1.
6. Add the time cost distribution of the candidate task Ti to the current load of the 1st processor in the candidate processor list, say Pf, f ∈ {1,2, ,m}, and evaluate the expected time cost μ'_{P_f} and its variance, V'_{P_f}, for the resultant time cost distribution of Pf.
7. If $\mu'_{P_r} + \sigma'_{P_r} \le \mu + 2\sigma$ then partitioning of Ti is not necessary.

a. Schedule the candidate task Ti on the processor Pf.
b. With equal constraint, discard the processor Pf from the scheduling process in next steps since it has its full load. Otherwise, replace the old mean and variance with the new mean and variance.

8. If $\mu'_{P_r} + \sigma'_{P_r} \succ \mu + 2\sigma$ the candidate task Ti must be partitioned.

   Add the kth split task to the current load of this processor. Other partitions are scheduled on the first k processors in the candidate processor list. Evaluate the expected time cost and variance for the results. Find the smallest k such that $\mu'_{P_r} + \sigma'_{P_r} \leq \mu + 2\sigma$. If such k exists, schedule the k partitions of Ti on the first k processors in the candidate processor list.

9. If $\mu'_{P_r} + \sigma'_{P_r} \succ \mu + 2\sigma$, make k =m (or the maximum number of partitions) and repeat step 8.
10. Update the expected time cost and the variance for the current loads of the selected processors. Discard the full loaded (or overloaded) processors from the scheduling process in next steps. Reorder the working processors in the candidate processor list.
11. increment i and repeat steps 6 to 9 until i=n. In this case, evaluate the expected time cost and the variance for the finishing time distribution of each processor in the system. Choose the finishing time distribution of the processor with the maximum expected time cost to be the suitable distribution for the finishing time of the schedule, FTS. If more than one distribution has the same maximum, choose the one of maximum variance.

We conducted a simulation study. In this study, we investigated in whether or not each processor contributes an equal computational workload. Therefore, the most appropriate metric to evaluate the merits of the algorithm is the probability of finishing time of the schedule, FTS, to meet the desired schedule time, TS, i.e., P(FTS ≤ TS). We assumed that TS=μ+2σ. Sine the FTS is given by the sum of several independent random variables, it follows approximately a normal distribution with mean μFTS and variance VFTS, according to the central limit theorem. Therefore, μFTS and VFTS, were used to find the required probability from the standard normal tables. We considered that the algorithm provides a well-balanced load, if P(FTS≤TS) is at least equal to 85%. A set of tasks, with different means and variances, was randomly generated. In each test case, we assumed that the number of tasks, n, is greater than the number of processors, m. The total number of tasks varies from 10 to 500, while the number of processors varies from 4 to 32. The maximum number of partitions, k, varies from 2 to the available number of processors. The worst case means and variances of a task Ti partitions when executed on k processors in parallel were evaluated by using the sub linear relations. Results show that the proposed algorithm provides better results than the algorithm without partitioning specially when the number of tasks and processors grow. We also found that the probability of finding a flat time distribution increases as the number of partitions increases.

No. of processors		50≤μ<100 10≤V<20	100≤μ<150 20≤V<15	150≤μ<300 30≤V<60	300≤μ<500 60≤V<100	500≤μ<750 100≤V<150
4	Without par.	80.2%	78.7%	75.3%	70.6%	65.2%
	With par.	96.1%	93.8%	90.6%	88.9%	87.1%
8	Without par.	77.7%	75.2%	71.9%	65.2%	61.3%
	With par.	91.3%	89.7%	86.4%	82.8%	81.4%
16	Without par.	70.4%	69.6%	66.2%	63.3%	57.7%
	With par.	88.2%	84.4%	82.6%	80.1%	78.8%
24	Without par.	68.9%	61.4%	60.1%	59.9%	55.6%
	With par.	84.4%	82.2%	80.8%	78.4%	77.6%
32	Without par.	65.2%	60.3%	59.1%	57.7%	51.9%
	With par.	83.6%	82.1%	80.4%	78.3%	77.1%

Effective scheduling strategies are essential to improve the execution time performance, throughput, and the processor utilization in multiprocessor systems. In this paper, we presented two heuristic scheduling algorithms for stochastic tasks with known time cost distributions. The second one assumes that tasks can be partitioned. The proposed algorithms can provide a nearly flat distribution of computational workload to all the available processors in the system.

5. References

[1] E.Y. Abdel Maksoud and R.A. Ammar "A Heuristic scheduling Algorithm for Stochastic Tasks in Distributed Multiprocessor Systems", Proc. ISCA 15th Int'l Conf. on Parallel Computing and Distributed Systems, Louisville, Kentucky, pp. 277-282, Sep. 19-21, 2002.

[2] B. Bataineh and B. Al-Asir, "Efficient Scheduling Algorithm for Divisible and Indivisible Tasks in Loosely Coupled Mulitprocessor Systems," Software Eng. J., pp. 13-18, 1994.

[3] Olivier Beaumont, Henri Casanova, Arnaud Legrand, Yves Robert, Yang Yang Scheduling Divisible Loads on Star and Tree Networks: Results and Open Problems, IEEE Trans. on Parallel and Distributed Sys. PP 207-218, March 2005.

[4] A. Dogan and F. Osgunger, "Matching and Scheduling Algorithms for Minimizing Execution Time and Failure Probability of Applications in Heterogeneous Comp." IEEE Trans. on Parallel and Distributed Systems, PP 308-323, June 2002.
[5] H. El-Rewini, T.G. Lewis, and H.H. Ali, Task Scheduling in Parallel and Distributed Systems. New Jersey: Prentice Hall, 1994.
[6] C. Guus, E. Boender, and H.E. Romeijn, "Stochastic Methods," Handbook of Global Optimization, R. Horst and P.M. Pardalos, eds., pp. 829-869. The Netherlands: Kluwer Academic Publishers, 1995.
[7] B.A. Shirazi, A.R. Hurson, and K.M. Kavi, "Scheduling and Load Balancing in Parallel and Distributed Systems," IEEE Computer Society Press, 1995.
[8] M.Y. Wu, "On Runtime Parallel Scheduling for Processor Load Balancing" IEEE Trans. Parallel and Distributed Systems, vol. 8, no. 2, pp. 173-186, Feb. 1997.
[9] C. Xu and F. Lau, Load Balancing in Parallel Computers. Boston: Kluwer Academic Publishers, 1997.

Brill Academic Publishers
P.O. Box 9000, 2300 PA Leiden,
The Netherlands

Lecture Series on Computer and Computational Sciences
Volume 2, 2005, pp. 6-10

Approximating by Szsz-Type Operators

A. Aral [1]

Mathematics Department,
Arts and Sciences Faculty,
Kırıkkale University,
Yahşihan, Kırıkkale 71100 Turkey

H. Erbay [2]

Mathematics Department,
Arts and Sciences Faculty,
Kırıkkale University,
Yahşihan, Kırıkkale 71100 Turkey

Received 5 April, 2005; accepted in revised form 16 April, 2005

Abstract: We introduce a new Szász-Type operators depending on weighted functions. We analyze approximation results of these operators on weighted space. Our numerical results are consistent with our theory.

Keywords: Szsz-Type operator, weighted space, weighted approximation, modulus of continuity.

Mathematics Subject Classification: 41A36

1 Introduction

Let $(L_n)_{n\geq 1}$ be a sequence of positive linear operators such that

$$\lim_{n\to\infty} \|L_n\varphi_i - \varphi_i\|_{C[a,\,b]} = 0, \;\; i = 0,\, 1,\, 2, \tag{1}$$

where $\varphi_i(x) = x^i$, $i = 0, 1, 2$. Then $(L_n)_{n\geq 1}$ converges to every continuous function f on $[a, b]$ by Korovkin theorem [6]. In other words, convergence on three functions may be extended to all functions which are continuous on $[a, b]$ and bounded on $\mathbb{R}$. The functions $\varphi_i(x)$, $i = 0, 1, 2$ are called the test functions. Similar result on approximation of functions by the sequence of positive linear operators may be given via the modulus of smoothness. For example in [2], let $(L_n)_{n\geq 1}$ be a sequence of positive linear operators on $C[0, 1]$ into itself and $\lambda_n = \max_{i=0,1,2} \|L_n\varphi_i - \varphi_i\|_{C[0,\,1]}$. Then for each $f \in C[0, 1]$

$$\|L_n f - f\|_{C[0,1]} \leq \kappa \left\{\lambda_n \|f\|_{C[0,1]} + \omega_2\left(f;\, \sqrt{\lambda_n}\right)\right\} \quad n = 1, 2, ... \tag{2}$$

with $\omega_2(\cdot;\delta)$, $0 < \delta < 1$, the second order of moduli of smoothness as measured in $\|\cdots\|_{C[0,1]}$ and here κ is a positive constant. Since $\lim_{n\to\infty} \lambda_n = 0$, by (1) we have $\lim_{n\to\infty} \omega_2(f;\, \sqrt{\lambda_n}) = 0$ for the uniformly continuous function on $[0, 1]$ (see [1]). Therefore the operators $L_n f$ converges to f on $C[0.1]$ by (2).

With respect to weight norm, the extensions of classical Korovkin's theorem for the continuous functions belonging to weighted space on unbounded intervals was carried out by Gadjiev [3] . He showed that for the sequence of positive linear operators $(L_n)_{n\geq 1}$ converging to f in weight norm, there must be a relation between the test functions and weighted functions. To explain why, we recall the weighted spaces $B_\rho(\mathbb{R})$ and $C_\rho(\mathbb{R})$ given in [3, 4].

For any weight function ρ defined on $\mathbb{R}$,

$$B_\rho(\mathbb{R}) := \{f : |f| < M_f \cdot \rho, \rho \geq 1 \text{ and } \rho \text{ unbounded}\}$$

and

$$C_\rho(\mathbb{R}) := \{f : f \in B_\rho \text{ and } f \text{ continous}\} .$$

We restrict our attention to $\mathbb{R}^+ := [0, \infty)$ rather than $\mathbb{R}$ and denote the norm of f belonging to $B_\rho(\mathbb{R}^+)$ by

$$\|f\|_\rho = \sup_{x\geq 0} \frac{|f(x)|}{\rho(x)}.$$

We also recall the weighted space

$$C_\rho^k(\mathbb{R}^+) := \left\{ f : f \in C_\rho(\mathbb{R}^+) \text{ and } \lim_{x\to\infty} \frac{f(x)}{\rho(x)} = k_f < \infty \right\}.$$

It is obvious that $C_\rho^k(\mathbb{R}^+) \subset C_\rho(\mathbb{R}^+) \subset B_\rho(\mathbb{R}^+)$.

Throughout the paper, we consider the weight function ρ with the assumptions:

(i) ρ is a continuously differentiable function on $\mathbb{R}^+$,

(ii) $\inf_{x\geq 0} \rho'(x) \geq 1$.

Lemma A 1 *$(L_n)_{n\geq 1}$ is the sequence of positive linear operators from $C_\rho^k(\mathbb{R}^+)$ to $B_\rho(\mathbb{R}^+)$ if and only if*

$$|L_n(\rho; x)| \leq K\rho(x),$$

where K is a positive constant.

Theorem A 1 *If the sequence of positive linear operators $L_n(f; x)$ acting from $C_\rho(\mathbb{R}_0)$ to $B_\rho(\mathbb{R}_0)$ satisfies the conditions:*

$$\lim_{n\to\infty} \|L_n(\rho^\nu; x) - \rho^\nu(x)\|_\rho = 0, \ \nu = 0, 1, 2 \tag{3}$$

then

$$\lim_{n\to\infty} \|L_n(f; x) - f(x)\|_\rho = 0$$

for any function $f \in C_\rho^0$

Note by Theorem A that, the weight function ρ is related with both the test functions and the approximated function f. From the properties of function f and the results of Theorem A we say that the weighted function ρ not only explains behavior of function f at infinity but also gives construction of the test functions in the weighted spaces.

In this paper, we consider the linear positive operators whose test functions are $1, \rho, \rho^2$ [5]. We analyze the approximation properties in weighted spaces. We give both direct approximation and also an estimate including modulus of continuity. Finally we test the theory that we develop by numerical examples.

2 Structure of the Szász-Type Operator

Let $(L_n)_{n\geq 1}$ a sequence of positive linear operators such that

$$\begin{aligned} L_n(f;x) &= \rho^2(x)\sum_{k=0}^{\infty}\frac{f\left(\frac{k}{n}\right)}{\rho^2\left(\frac{k}{n}\right)}e^{-nx}\frac{(nx)^k}{k!} \\ &= \rho^2(x)\,S_n\left(\frac{f(t)}{\rho^2(t)};\,x\right), \end{aligned}$$

where $S_n(f;\,x)=\sum\limits_{k=0}^{\infty} f\left(\frac{k}{n}\right)\,e^{-nx}\frac{(nx)^k}{k!}$ is the classical Szász operators, $n=1,2,...,$ and $x\in\mathbb{R}^+$.

One can easily show that

$$L_n(1,\,x)-1=\rho^2(x)\left[\sum_{k=0}^{\infty}\frac{1}{\rho^2\left(\frac{k}{n}\right)}e^{-nx}\frac{(nx)^k}{k!}-\frac{1}{\rho^2(x)}\right], \tag{4}$$

$$L_n(\rho,\,x)-\rho(x)=\rho^2(x)\left[\sum_{k=0}^{\infty}\frac{1}{\rho\left(\frac{k}{n}\right)}e^{-nx}\frac{(nx)^k}{k!}-\frac{1}{\rho(x)}\right], \tag{5}$$

$$L_n\left(\rho^2,\,x\right)-\rho^2(x)=0. \tag{6}$$

By taking $x\in[0,\;a]$, $a>0$, we let

$$\gamma_n=\|L_n(1)-1\|_{C[0,\;a]}, \tag{7}$$

$$\beta_n=\|L_n\rho-\rho\|_{C[0,\;a]}\,. \tag{8}$$

Theorem 1 *For any function $f\in C_\rho^k$ if $\lim\limits_{n\to\infty}\|S_n(\rho^\nu)-\rho^\nu\|_\rho=0$, $\nu=0,\,1,\,2$ then $\|L_n-f\|_{\rho^3}\to 0$ as $n\to\infty$.*

Theorem 2 *Suppose that the function f belongs to $C_\rho(\mathbb{R}^+)$. Then, for the sequences of positive linear operators $\{L_n\}$ the inequality*

$$\|L_n(f;\,x)-f(x)\|_{C[0,\,a]}\leq K\omega_{[0,\,a]}\left(f,\,(\gamma_n+\beta_n)^{1/2}\right),$$

holds. Here $\omega_{[0,\,a]}(f,\delta)$ is the modulus of continuity of finite interval $[0,\;a]$.

3 Numerical Examples

In this section we report some selected numerical examples that correct our theory. We run Matlab 6.5 on a personal computers to obtain results.

In both examples, the interval divided into 100 subintervals of equal widths. The graph of L_n for each n is obtined by plotting $L_n(f;\,x)$ in blue at each endpoints the subintervals. The figures shows the graphs of first 15 terms of L_n.

We also plotted $f(x)$ in red at these endpoint.

Note from figures 1 and 2 that the terms of L_n gradually approaches the graph of f.

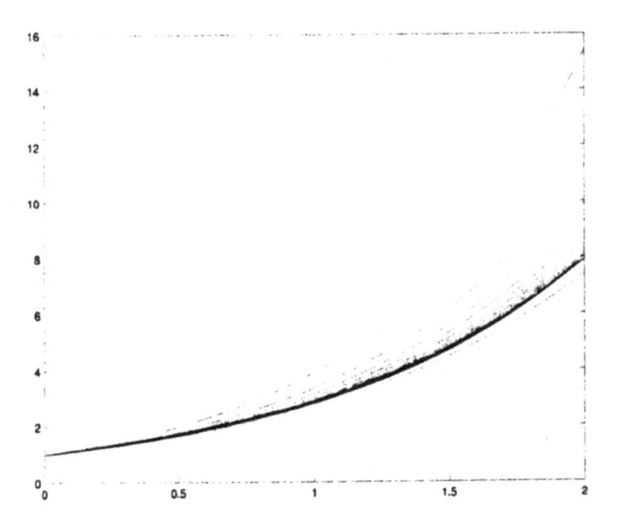

Figure 1: By example 1

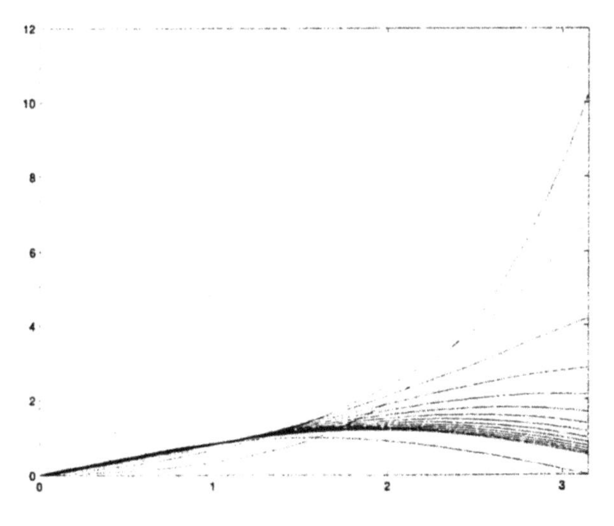

Figure 2: By example 2

References

[1] Z. Ditzian and V. Totik, *Moduli of Smoothness*, Springer Series in Computational Mathematics, Vol 9, Springer-Verlag, Berlin / Heidelberg / New York, 1987.

[2] G. Freud, On approximation by positive linear methods I, II, Studia Sci. Math. Hungar. 2(1967) pp. 63-66; 3(1968) pp. 365-370.

[3] A.D. Gadjiev, On P. P. Korovkin type theorems, Math. Zamet., 20(1976) pp. 781-786. Transl in Math Notes (1976) N5-6, (1978) pp. 995-998.

[4] A. D. Gadjiev, The convergence problem for a sequence of positive linear operators on bounded sets and theorems analogous to that of P. P. Korovkin. Soviet Math. Dokl., V.15, No:5, (1974) pp. 1433-1436.

[5] A. D. Gadjiev and A. Aral, *The Estimates of Approximation by Using a New Type of Weighted Modulus of Continuity*, submitted for publication.

[6] P. P. Korovkin, *Linear Operators and Approximation Theory*, Hindustan Publ. Corp. (Delhi, 1960.

Brill Academic Publishers
P.O. Box 9000, 2300 PA Leiden,
The Netherlands

Lecture Series on Computer and Computational Sciences
Volume 2, 2005, pp. 11-15

Implicit Versus Explicit Classification: Excerpts From a Concept Analysis of Classification

T.K. Bjelland[1]

Stord/Haugesund University College,
Bjornsonsgate 45, 5528 Haugesund, Norway

Received 28 February, 2005; accepted in revised form 16 March, 2005

Abstract: The purpose of this paper is to present some results from a concept analysis of classification. Classification in this respect is understood as the process of defining the key concepts in an application domain, as well as the vocabulary that results from the process. The analysis emphasizes the importance of membership conditions to define classes and identify class members, and suggests that classification is a prerequisite to the data modeling process. It is assumed that the vocabulary will be of valuable help to designers engaged in the design of conceptual data models, and to end-users and internal auditors when interpreting and validating conceptual data models.

*Keywords:*Classification, concept, class, intension, extension, membership condition

*Mathematics Subject Classification:*03E99

1 Introduction

Classification is generally considered as a fundamental abstraction mechanism and important modeling principle in conceptual modeling. Yet, in spite of its importance, the discipline seems to lack a unified account of the term. So, what does classification mean? This paper presents some results from a concept analysis of the term 'classification'. From a total of 288 documents selected for the study, 115 papers were randomly selected for analysis. Some papers considered to be classic were included, increasing the total sample size to n=143. Sorted by content the papers gave the following gross distribution:

Table 1: Gross distribution of papers sorted by content.

Topic	No. of papers
Cognitive aspects of classification	16
Representational aspects, including representational languages and modeling approaches	66
Practical and theoretical aspects, including principles and techniques, taxonomies and typologies	31
Schema integration	30

[1]Corresponding author. E-mail: tor.bjelland@hsh.no

In this paper emphasis is placed on the analysis of representational aspects, and special attention is given to the notion of sets and membership conditions in association with classification. The analysis shows that membership conditions are almost unheard of in conceptual modeling, and that classification is performed as a kind of intuitive practice, where classes are named but not properly defined. As a consequence, important distinctions between intensional and extensional aspects of classes are overlooked, making it hard to make sense of the key concepts used in conceptual modeling.

2 Representational Aspects of Classification

One of the earliest proposals for semantic data models was Abrial's paper on Data Semantics in 1974, [1]. The model consists of objects, categories, and binary relations between pairs of objects that belong to the categories. Although categories are used to represent sets of objects, there is no direct mention of classification, concepts, or membership conditions. In connection with an example, it is said that various objects are intuitively organized into different categories. This notion of 'intuitive classification' is also used by [4] in one of his earlier writings where its is said that we recognize and classify 'intuitively' entities into sets. In a sense, the term 'intuitive classification' seems to name a method of classification where data modeling is guided by rules on how to *discover* existing classes. According to [11] the object-oriented and semantic data modeling literature also offers advice on identifying classes. However, this advice is usually very general, such as identifying tangible things, roles and events, and is not of much help in determining whether a selected set of classes is appropriate. This idea of identifying classes is also commented by [3] who claims that most modeling methods implicitly assume that object classes do exist in the world, waiting to be discovered by the data modeler. The author mentions three reasons for this:

- First, many database systems are replacing older systems in which classes are already defined.
- Second, in common speech, we frequently use general terms to refer to individual objects suggesting that many objects belong to a single class.
- Third, some classes, with which we have first hand physical experience, seem so natural that it is hard to see what other way the individual in that class could be identified.

This last reason appears to have very much in common with the so called 'intuitive types' in archaeology, which, according to [2], jump at the classifiers in the form of intuitive gestalts. Something that speaks against 'intuitive classification' is the fact that several prominent writers in the field take a predicate view on classification. In Chen's famous paper from 1976, [5], a test-predicate is explicitly mentioned for both entity sets and value sets. Entities are said to be classified into different entity sets and a test predicate is associated with each entity set to test whether an entity belongs to it. The test predicate is a property and can be found among the properties that the entities in an entity set have in common. Similarly, it is said that values are classified into value sets and that a predicate is associated with each value set to test whether a value belongs to it. In addition to the explicit mentioning of a test predicate, Chen also makes a very important distinction that is generally not recognized in conceptual modeling: a) that entities in an entity set have properties common to the other entities in the entity set, and b) that, among these properties, there is also a test predicate. Usually, descriptions of entity sets and similar constructs contain a), but not b). The distinction that Chen makes suggests a distinction between definitional properties and descriptive properties. While the definitional properties, which constitute the test predicate, necessarily must be invariant and the same for each and every entity in the entity-set, the descriptive properties will normally be variant and dissimilar for most entities. Take for instance a class of male students. While the defining predicate for the class requires every student to be of the same sex = 'male', properties such as SSN, name, date of birth, address, phone, etc., will differ for each student, and some of the properties will even change over time. The distinction between defining and descriptive properties is not directly noticed by [12], although the authors stress the importance of classification in designing database models, but use the term generalization instead of classification. In addition, [12] describe and demonstrate a method for representing a generic hierarchy as a hierarchy of relations. Their ideas are included and further detailed in a paper by Codd [6] where he proposes extensions to the relational model in order to capture more of the semantics in a database. Codd distinguishes between two extensional aspects of generalization: instantiation and subtype. Both are forms of specialization, and their reverses are forms of generalization. The extensional counterpart of instantiation is set membership, while that of subtype is set inclusion. In a paper by [10], the two extensional aspects have been formalized and termed classification and generalization, respectively. Codd also recognizes that entities may belong to (or be described by) several types, and that there is a need for unique and permanent identifiers to keep track of those entities for which there may be several descriptions. Here, Codd identifies the descriptive aspect of types, but he fails to recognize any definitional aspects. According to Codd, classification, or generalization by membership, as he calls it, is taken care of by E-Relations. An E-Relation is a unary relation where entity types are represented by a name only. The general idea is that the predicate or membership condition should be reflected by the name. Again, this is clearly an example of 'intuitive classification'. On the other hand, when it comes to generalization by inclusion, Codd distinguishes between unconditional and conditional generalization. Both kinds are represented by a triple relation (SUB:m, SUP:n, PER:p), where m represents the subtype, n, its supertype, and p the predicate. The strange thing here is that predicates are explicitly used to control membership in subtypes, but no predicates are stated for the supertypes, or types that do not participate in generalization hierarchies. Another well known paper from this period is [9], where the authors propose a model to formally specify the meaning of a database. With respect to classification, they propose that a class should have an optional textual class description describ[ing] the meaning and contents of the class. A class description should be used to describe the specific nature of the entities that constitute the class and to indicate their significance and role in the application environment. Although the 'class description' as they call it is optional, it is a very clear request for adding a membership condition to a class. In addition, and very similar to Codd, they propose predicate-defined subclasses, where a subclass S is defined by specifying a class C and a predicate P on the members of C. The subclass S is said to consist of just those members of C that satisfy P. The author in [10] make an effort to formalize the key concepts of object-oriented analysis. The important thing to notice here is the authors' claim that some concepts mean different things in different contexts such as analysis and design. Because of this, a concept for instance, is called a type in analysis and a class in design. Types define a problem, while classes represent a solution to the problem. Therefore, analysts are concerned with terminology and definitions in order to define the problem, while designers are concerned with efficient storage structures and inheritance in order to design an efficient solution. The terminological aspect is also very central in more recent AI-research on ontology building and use. According to [7], one of the main motivations for this research is the possibility of knowledge sharing and reuse across different applications. To achieve this, the applications must commit to the definitions in a common ontology. An ontology in this respect defines the terminology of a domain of knowledge, [13], and the commitment, which is known as 'ontological commitment' is defined as an agreement to use a shared vocabulary in a coherent and consistent manner [8]. This has much in common with the goals of conceptual modeling, though the focus on vocabulary and terminology is less emphasized.

3 Conclusion

This mix of statements from pioneers as well as more recent writers in the field, indicates that there exist at least two major views on classification. According to the first view, classification relies heavily on intuition. There are no formal rules, except for some simple heuristics, such as looking for people, things, roles, interactions, and places. As soon as the types are found, the relevant objects are expected to fall into their respective types automatically, so identification of members is not an issue. In this respect, classification can be viewed as a direct relation between a type and its associated objects, without any further references to mental concepts and terminology. The other view is more complex, and involves the interplay among cognitive, linguistic, and ontological elements. Here, classification has very much in common with terminology, where mental concepts are concretized and formally defined before they are arranged in a system of concepts. In accordance with this view, a mental concept is understood as an abstract conceptual entity that is characterized by an intension and an extension. In order for mental concepts to be shared and communicated, concepts must be defined. This process of defining concepts, and finding their correct position in a hierarchy of definitions, is called classification. As a process, classification produces a terminology that is suited to name the concepts that make up the universe of discourse. Hence, classification may be regarded as a prerequisite to the modeling task, since terms cannot be related to form a concept system unless one knows their definitions. Similarly, one may say that classification is a prerequisite to identification. One cannot identify something by a name alone, unless one knows how to apply the name correctly. Finally. we may say that identification is a prerequisite to description. Nothing can be described before it has been identified. I cannot describe something as a flower unless I have it before me and it has been recognized as a flower.

Acknowledgment

The author wishes to thank Professor Sylvia B. Encheva for valuable hints and suggestions, and Paul Glenn for his careful reading of the manuscript.

References

[1] Abrial, J.: Data Semantics. In: Klimbic and Koffeman (Eds.): Data Base Management North-Holland Publishing Company 1 59 (1974).

[2] Adams, W.Y.: Archaeological classification: theory versus practice, Antiquity 61, 40-56, (1988).

[3] J. Artz, A crash course on metaphysics for the database designer, Journal of Database Management, 8(4), 25 30, (1997).

[4] Bubenko J. Jr.: IAM: An Inferential Abstract Modeling Approach to the Design of Conceptual Schema. In: D.C.P. Smith (Ed.): Proceedings of the 1977 ACM Sigmod Conference on Management of Data. Canada (1977).

[5] Chen, P.P.: The Entity-Relationship Model - Towards a Unified View of Data. ACM Transactions on Database Systems. 4(4) (1979).

[6] Codd, E.: Extending the Database Relational Model to Capture More Meaning. ACM Transactions on Database Systems, 4(4), (1979).

[7] Guarino, N.: Understanding, building and using ontologies. International Journal on Human-Computer-Studies, 46, 293 310, (1997).

[8] Gruber, T. R.: Toward principles for the design of ontologies used for knowledge sharing. International Journal on Human-Computer Studies, 43, 907 928, (1995).

[9] Hammer, M., McLeod, D.: Database Description with SDM: A Semantic Data Model. ACM Transactions on Database Systems, 6(4), (1981).

[10] Odell, O.J., Ramackers, G.: Toward a Formalization of OO Analysis. Downloaded from http://www.quoininc/JOarticle9707.html

[11] Parsons, J., Wand, Y.: Choosing Classes In Conceptual Modeling, Communications of the ACM, 40(6), 63 69, June 1997.

[12] Smith, J.M., Smith, D.C.P.: Database Abstractions: Aggregation and Generalization. ACM Transactions on Database Systems, 2(2), 105 133, (1977).

[13] Waterson, A., Preece, A.: Verifying ontological commitment in knowledge-based systems, Knowledge-Based Systems 12, 45 54, (1999).

Brill Academic Publishers
P.O. Box 9000, 2300 PA Leiden,
The Netherlands

Lecture Series on Computer and Computational Sciences
Volume 2, 2005, pp. 16-19

Smoothing Finite-Element Method for the Reconstruction of Surface from 3D Scattered Data

Zhong-Yi Cai[1], Ming-Zhe Li

Roll Forging Research Institute,
Jilin University (Nanling Campus)
5988 Renmin Street, Changchun, 130025, P.R.China

Received 10 April, 2005; accepted in revised form 25 April, 2005

Abstract: This paper presents a numerical method for reconstructing freeform surface from 3D scattered measured data. A positive definite functional is formulated by introducing a smoothing factor and minimized by use of finite-element to best-fit the scattered data, and thereby C^0, C^1 or C^2 continuity surfaces are represented. The applicability of the method is illustrated by reconstructing surfaces with eight-node isoparametric element, 18 d.o.f. triangular element and second-order Hermite element, respectively. Effects of the errors and the amounts of input data on the reconstructed results are investigated.

Keywords: finite element; smoothing; surface reconstruction; optimal approximation

Reconstructing a smooth surface from scattered data is of notable importance to the industry. Such a process has been studied extensively [1-10]. The scattered data are typically measured in a region and contain measuring errors. It is unwise to reconstruct a surface, which agrees strictly with measured data at input locations. Particularly if derivatives are sought, it is normally advantageous to formulate a smoothing approximating representation of the measured data.
A so-called smoothing finite-element method (SFEM) is presented here to reconstruct smooth surface from discrete measured data. The method combines the energy smoothing capability with the finite-element concept and reliable first-order derivatives even acceptable second-order derivatives can be predicted since the influence of noise in input data is eliminated effectively.

1. Description of the concept and formulations

In the region of interest Ω, a set of measured data $\hat{f}(x_i)$, $i = 1,2,\cdots,L$ are given. We use an energy model — a flexed plate loaded by stretched springs attached at the data locations to obtain the smooth surface. The total elastic energy in springs are expressed by

$$E_1 = \frac{1}{2}\sum_{i=1}^{L} C_i[f(x_i) - \hat{f}(x_i)]^2 \tag{1}$$

where C_i is the spring constant at data location i. We take $C_i = (\delta f_i)^{-2}$ here, δf_i is the estimated errors of $\hat{f}(x_i)$. Obviously, to best fitting the given data E_1 should be approximate to $L/2$.
Based on the elastic-plate analogy [11,12], the elastic strain energy is commonly defined by curvatures of the surface as follow

$$E_2 = \frac{1}{2}\int_{\Omega}\sum_{r=1}^{2}\sum_{s=1}^{2}\left(\frac{\partial^2 f(x)}{\partial x_r \partial x_s}\right)^2 \mathrm{d}\Omega \tag{2}$$

where x_r and x_s are the rth and sth coordinates of the position vector x in Ω, respectively.
By introducing a Lagrangian multiplier λ and a smoothing factor ω, leads to the following energy functional

[1] Corresponding author. E-mail. czy@jlu.edu.cn

$$E(f,\lambda)=\frac{1}{2}\int_{\Omega}\sum_{r=1}^{2}\sum_{s=1}^{2}\left(\frac{\partial^2 f(x)}{\partial x_r \partial y_s}\right)^2 \mathrm{d}\Omega+\frac{\lambda}{2}\left\{\sum_{i=1}^{L}\left[\frac{f(x_i)-\hat{f}(x_i)}{\delta f_i}\right]^2-\omega L\right\} \tag{3}$$

Thus, to obtain the optimal smooth surface, we have to look for the minimum of the above functional. If the measured data come from a curve, the functional reduces to essentially a one-dimensional problem. Reinsch [13] obtained a smooth cubic spline solution of Eq.(3) from the Euler-Lagrange equation of the functional, and the smooth curve can be constructed based on this solution. It is difficult to obtain exact extreme solution for the two- or three-dimensional problem. That difficulty is overcome here by seeking an approximate solution using finite element.

We divide region Ω into m finite-elements and n nodes. Within any individual element, f is then interpolated by $f = NF_e$, where F_e is the vector of nodal parameters in an element and N the shape function matrix. Minimization of Eq. (3) respect to all the unknown nodal parameters in Ω and Lagrange multiplier leads to following equations

$$\begin{cases}(\lambda^{-1}K_1+K_2)F=P\\ F^{\mathrm{T}}K_2F-2F^{\mathrm{T}}P+V^{\mathrm{T}}V=\omega L\end{cases} \tag{4}$$

here, K_1, K_2, P, F and V are the assembly of K_1^e, K_2^e, P_e, F_e and V_e over all m elements in region Ω, respectively, and

$$K_1^e=\int_{\Omega_e}\sum_{r=1}^{2}\sum_{s=1}^{2}\frac{\partial^2 N^{\mathrm{T}}}{\partial x_r \partial x_s}\frac{\partial^2 N}{\partial x_r \partial x_s}\mathrm{d}\Omega,\quad K_2^e=\tilde{N}^{\mathrm{T}}\Delta^{-2}\tilde{N},\quad P_e=\tilde{N}^{\mathrm{T}}\Delta^{-1}V_e \tag{5}$$

where $\tilde{N}=[\tilde{N}_1\ \ \tilde{N}_2\ \ \cdots\ \ \tilde{N}_{nc}]$ and $\tilde{N}_i=[N_i(x_1)\ \ N_i(x_2)\ \ \cdots\ \ N_i(x_{le})]^{\mathrm{T}}$; $V_e=\Delta^{-1}\hat{F}_e$ and $\Delta=\mathbf{Diag}[\delta f_1\ \ \delta f_2\ \ \cdots\ \ \delta f_{lc}]$, $\hat{F}_e=\{\hat{f}(x_1)\ \ \hat{f}(x_2)\ \ \cdots\ \ \hat{f}(x_{le})\}$, l_e is the number of data locations in element e .

Expression (4) is a nonlinear equation system with unknowns F and Lagrangian multiplier λ, it must be solved by an iterative procedure. To speed up the iterative process, let $\lambda=10^{\beta}$. By Newton-Raphson method, we have the following iterative scheme:

$$\begin{cases}F_k=(10^{-\beta_k}K_1+K_2)^{-1}P\\ \beta_{k+1}=\beta_k+\dfrac{10^{2\beta_k}}{2\ln 10}\dfrac{F_k^{\mathrm{T}}K_2F_k-2F_k^{\mathrm{T}}P+V^{\mathrm{T}}V-\omega L}{F_k^{\mathrm{T}}K_1^{\mathrm{T}}(10^{-\beta_k}K_1+K_2)^{-1}K_1F_k}\end{cases} \tag{6}$$

Two controlling factors are presented in surface reconstruction procedure. δf_i is an estimate of error associated with the various measured data. This value may or may not be the same for all data. The smoothing factor ω is used to obtain the surface that is best considered to approximate the measured data. As ω decreases in value, the Lagrangain multiplier λ gets large. When $\lambda\to\infty$, Eq.(3) degenerates to a pure least-square fitting of measured data ($(\delta f_i)^{-2}$ is considered as a weight factor), and nonlinear equation (6) degenerates to a linear equation

$$K_2F=P \tag{7}$$

The surface reconstructed in such case, while providing a reasonable fitting to the measured data, may contain excessive peaks and valleys. As ω increases in value, λ gets small and the curvature portion of Eq.(3) begins to dominate, a smooth surface is then yielded. At very small values of λ, the formulation tends to produce a least-square fitting of a plane to the measured data.

2. Numerical experiments and application example

C^0, C^1 and C^2 continuity surface can be constructed by eight-node isoparametric element, 18 d.o.f. triangular element and second-order Hermite element, respectively. The interpolating functions of

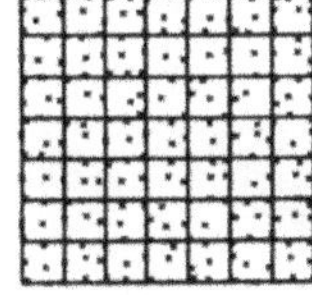

a. Input data and FE mesh

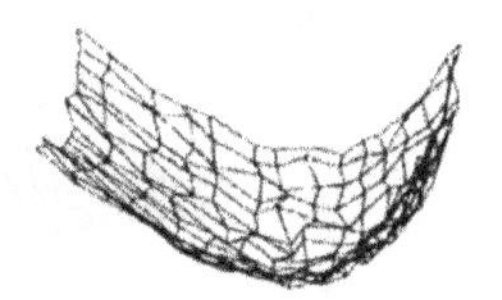

b. Original input data

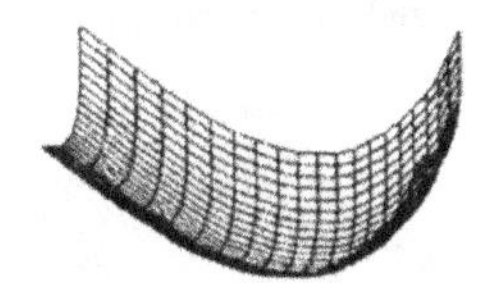

c. Surface reconstructed by SFEM

Fig. 1 A surface reconstructed by eight-node elements

eight-node isoparametric finite element are based on incomplete cubic polynomial. Fig.1 is a surface reconstructed by eight-node elements. The 225 measured data are scattered in a square region (Fig. 1a). 7×7 elements was divided,and input data in each element are 4~8. Fig.1b is the surface of original input data. Set $\omega = 0.9$, $\delta f_i = 0.3895$ and $\beta_0 = 0$, the convergent solution is $\beta = -2.305$. The surface reconstructed by SFEM is given in Fig.1c. It shows that SFEM provide a very smoothed surface.

The shape functions of the second-order Hermite elements are set up based on the Hermite interpolation polynomial. Six quantities are evaluated at each node

$$F_i = [f_i \quad f_{i,x} \quad f_{i,y} \quad f_{i,xx} \quad f_{i,xy} \quad f_{i,yy}]^T \tag{8}$$

$$N_i(\xi,\eta) = \lfloor H_i^{(0)}(\xi)G_i^{(0)}(\eta) \quad H_i^{(1)}(\xi)G_i^{(0)}(\eta) \quad H_i^{(0)}(\xi)G_i^{(1)}(\eta) \quad H_i^{(2)}(\xi)G_i^{(0)}(\eta) \quad H_i^{(1)}(\xi)G_i^{(1)}(\eta) \quad H_i^{(0)}(\xi)G_i^{(2)}(\eta)\rfloor \tag{9}$$

where

$$\begin{cases} H_i^{(0)}(\xi) = -\frac{1}{16}(3\xi^2+9\xi+8)(\xi-1)^3 \\ H_i^{(1)}(\xi) = -\frac{1}{16}(\xi+1)(5+3\xi)(\xi-1)^3 \\ H_i^{(2)}(\xi) = -\frac{1}{16}(\xi+1)^2(\xi-1)^3 \end{cases} (i=1,4), \quad \begin{cases} H_i^{(0)}(\xi) = \frac{1}{16}(3\xi^2-9\xi+8)(\xi+1)^3 \\ H_i^{(1)}(\xi) = \frac{1}{16}(\xi-1)(5-3\xi)(\xi+1)^3 \\ H_i^{(2)}(\xi) = \frac{1}{16}(\xi-1)^2(\xi+1)^3 \end{cases} (i=2,3) \tag{10}$$

$$\begin{cases} G_i^{(0)}(\eta) = -\frac{1}{16}(3\eta^2+9\eta+8)(\eta-1)^3 \\ G_i^{(1)}(\eta) = -\frac{1}{16}(\eta+1)(5+3\eta)(\eta-1)^3 \\ G_i^{(2)}(\eta) = -\frac{1}{16}(\eta+1)^2(\eta-1)^3 \end{cases} (i=1,2), \quad \begin{cases} G_i^{(0)}(\eta) = \frac{1}{16}(3\eta^2-9\eta+8)(\eta+1)^3 \\ G_i^{(1)}(\eta) = \frac{1}{16}(\eta-1)(5-3\eta)(\eta+1)^3 \\ G_i^{(2)}(\eta) = \frac{1}{16}(\eta-1)^2(\eta+1)^3 \end{cases} (i=3,4) \tag{11}$$

ξ and η denote the local coordinates of an element.

Fig.2 is a surface reconstructed by Hermite elements. The region was divided into 8×8 elements and 400 measured data are scattered (Fig. 2a) and described by Fig.2b, input data in each element are 4~9. Set $\omega = 1$, $\beta_0 = 0$ and $\delta f_i = 0.0569$, the convergent solution is $\beta = -0.6895$. Fig. 2c shows a very smoothed surface constructed by SFEM, while Fig. 2d shows the surface constructed by pure least-square fit (Eq.7) contains excessive peaks and valleys.

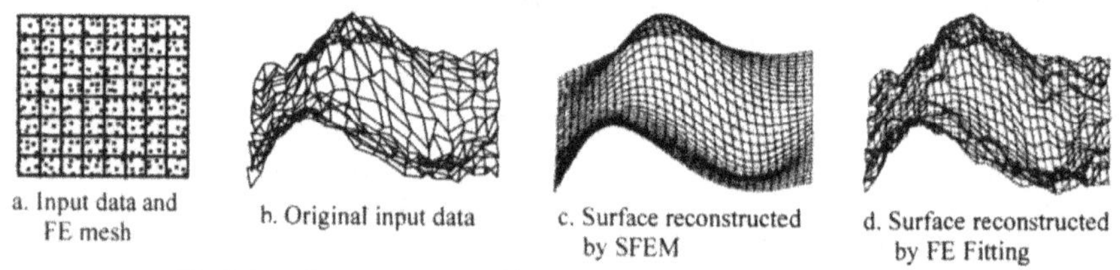

a. Input data and FE mesh b. Original input data c. Surface reconstructed by SFEM d. Surface reconstructed by FE Fitting

Fig. 2 A surface reconstructed by second-order Hermite finite-elements

The proposed method was applied to the duplication of a saddle-shaped part. The function of the objective surface of the part is $z = 30 + x^2/900 - y^2/900$ mm. Fig.3a is the 30×22 measured data obtained by coordinate measuring machine, and the surface was reconstructed by 18 degree-of-freedom triangular elements. A separate quintic interpolating function is assumed in each element, six quantities of Eq. (8) are evaluated at each of the three node points. In this element, $\partial f/\partial x$ and $\partial f/\partial y$ is continuous across common boundaries of contiguous elements, and the surface reconstructed is C^1 continuous in full field.

The measured region (400mm × 320mm) was divided by 160 triangular elements, values of $\omega = 1$, $\beta_0 = 0$ were used, after iterative process, the convergent solution is $\beta = 1.41$. Fig.3b shows the surface reconstructed by SFEM. Fig.4a and Fig.4b show the shape errors of measured data (Fig.3a) and reconstructed surface (Fig.3b), respectively. The mean error of z-coordinate of measured data is 0.904mm and the maximum error is 3.616mm, and the mean error of the reconstructed surface is

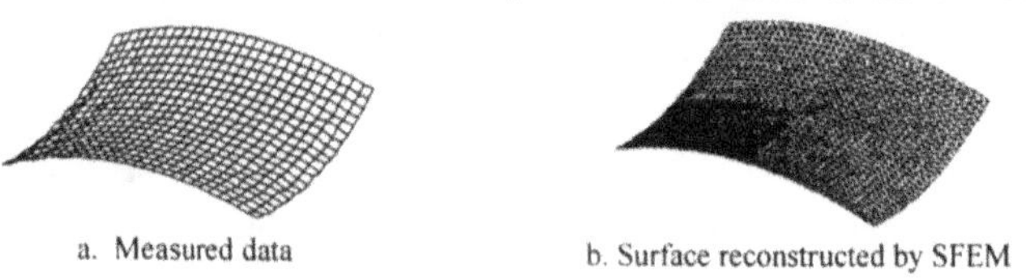

a. Measured data b. Surface reconstructed by SFEM

Fig. 3 The reconstructed results of saddle-shaped part

0.386mm and the maximum is 2.074mm. It can be seen that the reconstructed surface is of high approximating precision and good smoothness.

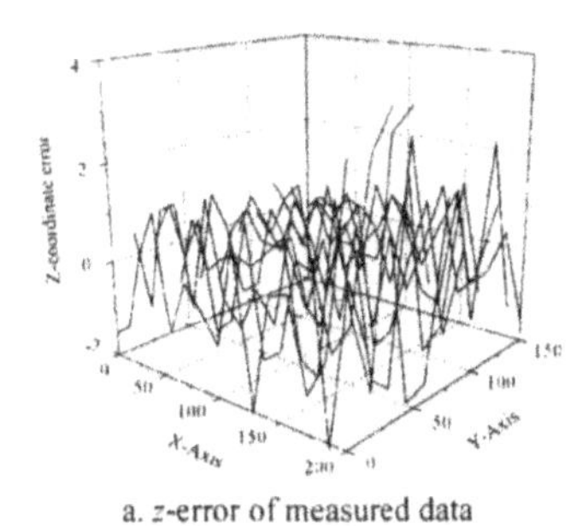

a. z-error of measured data

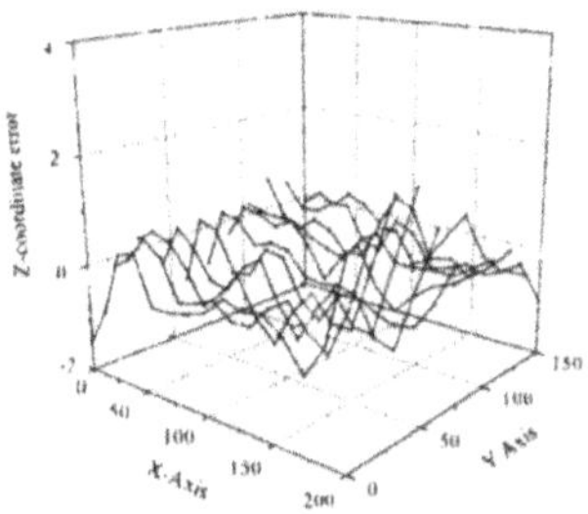

b. z-error of reconstructed surface

Fig. 4 The errors distribution

3. Concluding remarks

A smoothing finite-element method has been developed for the reconstruction of free-form surface from scattered measured data. The reliability of the method was assessed by numerical experiment and the applicability of the numerical method was additionally examined by reconstructing freeform surfaces with eight-node isoparametric finite element, 18 d.o.f. triangular finite-element and second-order Hermite finite- element, respectively.

The surface reconstructed by presented method is of high approximating precision and good smoothness. Among the three kinds of elements mentioned in this paper, Hermite elements produce the best results but such elements are merely applicable in rectangular region due to the limitation of element geometry. Although not so good results as those of Hermite element can yield, eight-node and 18 d.o.f. elements enable arbitrary geometric region to be fitted accurately.

Acknowledgments

This work was supported by National Science Foundation of China (50275063). The author wishes to thank the anonymous referees for their careful reading of the manuscript and their fruitful comments and suggestions.

References

[1] H. Hoppe, T. DeRose, T. Duchamp, J. McDonald, W. Stuetzle, Surface reconstruction from unorganized points, ACM SIGGRAPH Computer Graphics, 26(2), 71-78(1992)

[2] W. Ma and J. P. Kruth. Parameterization of randomly measured points for least squares fitting of B-spline curves and surfaces. *Computer-Aided Design*, 27(9), 663-675(1995)

[3] D. F. Rogers and N. G. Fog. Constrained B-spline curve and surface fitting. *Computer-Aided Design*, 21(10), 641-648(1989)

[4] B. Sarkar and C. H. Menq. Smooth-surface approximation and reverse engineering. *Computer-Aided Design*, 23(9), 623-628(1991)

[5] R.Franke. Scattered data interpolation: Tests of some methods. *Mathematics of computation*, 38,181-200(1982)

[6] J. Peters. Local smooth surface interpolation: a classification. *Computer Aided Geometric Design*, 7(1-4), 191-195(1990)

[7] T. N. T. Goodman and H. B. Said. A C^1 triangular interpolate suitable for scattered data interpolation. *Communications in Applied Numerical Methods*, 7, 479-485(1991)

[8] T. J. Wang A C^2 – quintic spline interpolation scheme on triangulation. *Computer-Aided Geometric Design*, 9 (5), 379-386(1992)

[9] G. Nielson. A method for interpolating scattered data based upon a minimum norm network. *Mathematics of computation*, 40(161), 253-271(1983)

[10] H. Pottmann. Interpolation on surfaces using minimum norm networks. *Computer-Aided Geometric Design*, 9(1), 51-67(1992)

[11] G. Celniker and D. Gossard. Deformable curve and surface finite elements for free-form shape design. *ACM SIGGRAPH Computer graphics*, 25(4), 257-266(1991)

[12] D. Terzopoulos, Platt J and Barr A *et al.* Elastically deformable models. *ACM SIGGRAPH* Computer Graphics, 21(4), 205-214 (1987)

[13] C. H. Reinsch. Smoothing by spline functions, *Numerische Mathematik*, 10,177-183(1967)

Brill Academic Publishers
P.O. Box 9000, 2300 PA Leiden,
The Netherlands

Lecture Series on Computer and Computational Sciences
Volume 2, 2005, pp. 20-23

An Online Collaborative Stock Management Mechanism for Multiple E-retailers

T. Chen[1] Guei-Sian Peng[2] Hsin-Chieh Wu[3]

[1,2]Department of Industrial Engineering and System Management,
Feng Chia University,
Taichung, Taiwan 407, ROC
[3]Department of Industrial Engineering and Management,
Chaoyang University of Technology
Taichung County, Taiwan 413, ROC

Received 27 March, 2005; accepted in revised form 7 April, 2005

Abstract: An e-retailer collaborates with other e-retailers to satisfy customer requirements in various ways. For example, in affiliate marketing the affiliate (or associate) e-retailer promotes the products and services of the merchant e-retailer on the affiliate's website for a commission. Another collaboration type is the cooperative exchanging mechanism proposed by Ito et al. in which an e-retailer can online exchange for a product if he does not have enough of the product for sale, with the aid of software agents. In this study, an online collaborative stock management mechanism for multiple e-retailers is proposed, in which a product can be simultaneously for sale on multiple websites, and the stocks of the product on these websites are monitored, virtually gathered together, and shared among websites.

Keywords: E-retailer; Stock management; Electronic Commerce

1. Introduction

Zwass [1] identified the five principal aspects of e-commerce development as commerce, collaboration, communication, connection, and computation. These aspects can be exploited to find innovational opportunities to organize and address marketplaces, offer innovative products, collaborate with business partners, transform business processes, and organize the delivery of information-system services. From another viewpoint, e-commerce is usually classified into business-to-business (B2B), business-to-consumer (B2C), consumer-to-consumer (C2C), consumer-to-business (C2B), non-profit-making, and intra-organizational categories, according to the participants involved [2]. If we map the innovational opportunities to these categories, then a matrix showing the innovational opportunities in these e-commerce categories can be constructed in Table 1. There are thirty cells and among them we focuses on the shaded cell, which indicates that innovational opportunities can be found to enhance the collaboration among business partners in B2C e-commerce, in which a business (the so-called e-retailer) sells products or services to final customers online.

Table 1: Innovative opportunities in e-commerce categories.

Innovative opportunities \ Categories	B2B	**B2C**	C2C	C2B	Non-profit-making	Intra-organizational
Organize and address marketplaces						
Offer innovative products						
Collaborate with business partners						
Transform business processes						
Organize the delivery of information-system services						

[1] Corresponding author. E-mail: tolychen@ms37.hinet.net

An e-retailer collaborates with its partners to satisfy customer requirements in various ways. For example, in affiliate marketing the partner of an e-retailer is another e-retailer, and the latter (the affiliate or associate) promotes the products and services of the former (the merchant) on the affiliate's website for a commission. In addition, the affiliate's site content can therefore be more fresh and dynamic. Many famous websites have provided affiliate marketing programs, e.g. the Amazon.com Associates program and the Yahoo Affiliates program. Similarly, Ito et al. [3] proposed a cooperative exchanging mechanism in which an e-retailer can online exchange for a product if he does not have enough of the product for sale, with the aid of software agents. In this paper, an online collaborative stock management (OCSM) mechanism is proposed for the same purpose. In the OCSM mechanism, the stocks of a product on multiple websites are monitored, virtually aggregated, and shared among these websites. Unlike an affiliate marketing program in which the merchant website determines the price, is actually responsible for the delivery of the product, and occupies the total profit, every website in the OCSM mechanism can determine a different price for the same product and makes a different profit. On the other hand, Ito et al.'s model considers only exchanging products between two websites. However, in the OCSM mechanism, the stocks of a product on multiple websites can be shared online among these websites. At last, to demonstrate the applicability of the proposed OCSM mechanism, an experimental system has been constructed, and then the advantages of the mechanism can be concluded.

2. The OCSM Mechanism

An OCSM mechanism is proposed in this study with the objective to increase the possibility of selling more products for e-retailers joining the mechanism. A product sold with the mechanism has the following characteristics:

(1) The product can be sold from any participating website.

(2) The product can be simultaneously for sale on all participating websites.

(3) The stock of the product on any participating website is limited.

With the OCSM mechanism, the stocks of the product on all participating websites are monitored, virtually aggregated, and shared among these websites. As a result,

(1) It becomes possible for a website to sell a product more than the current stock quantity. Here we assume that the price for trading a product among e-retailers has been negotiated in advance.

(2) The more a product has been sold from a website, the less the product can still be sold from the other websites, which results in a need for coordination that will be carried out with the OCSM mechanism.

There are three roles in the proposed mechanism:

(1) The original seller of a product, which is the website that sells the product and inputs the basic data of the product into the system database.

(2) The collaborative seller of the product: In addition to the original seller of the product, other websites joining the collaborative sale of the product afterward are called the collaborative sellers of the product.

(3) The system manager: The system manager is a software agent composed of three parts - a register dealing with the registration of websites and products, a stock monitor continuously recording and aggregating the stock levels on all participating websites, and a request sender sending requests to all participating websites to get or update the current stock levels.

The most important consideration in constructing the OCSM mechanism is not to disturb the normal operations of participating websites. The operation procedure of the OCSM mechanism is as follows (see Figure 1):

(1) An e-retailer goes to the system register to register its basic data.

(2) After checking the validity of the registration data, the e-retailer is informed of a set of username and password to register new products that will be sold with the mechanism. The e-retailer can also join the collaborative sale of other products that have been involved in the mechanism.

(3) The e-retailer has to download a stock reader and a stock updater that are simple script files (see Figure 2) used only to read and update the stock levels of products involved in the OCSM mechanism, respectively. The default values of parameters in the script files have to be modified.

(4) The request sender of the system manager sends a request to run the stock reader on each participating website every a small time interval, and the stock reader will respond the current stock levels of products involved in the OCSM mechanism on the website. Then the stock levels are aggregated to obtain the total stock levels (the total amounts that still can be sold online). The calculation process is explained as follows. Assume there are totally m products involved in the mechanism and denoted by P_i, $i = 1 \sim m$. There are n websites selling product P_i and indicated by W_{ij}, $j = 1 \sim n$. The current stock level of product P_i on website W_{ij} is S_{ij}. The previously obtained

total stock level of product P_i is PTS_i. Then the current total stock level of product P_i can be obtained as:

$$CTS_i = PTS_i - \sum_{j=1}^{n} (PTS_i - S_{ij}). \tag{1}$$

For example, assume product P_1 are for sale on two websites W_{11} and W_{12} collaboratively. The previously obtained total stock level of P_1 is 80, which represents the quantity of P_1 that still can be sold online since the last time. Assume the responded current stock levels of product P_1 on these two websites are 75 and 60, respectively. Then the current total stock level of product P_1 is 80 - (80 - 75) - (80 - 60) = 55.

(5) The request sender of the system manager sends a request to run the stock updater on each participating website to update the stock level of product P_i on the website to the current total stock level CTS_i. The PTS_i of product P_i is also updated to the same value.

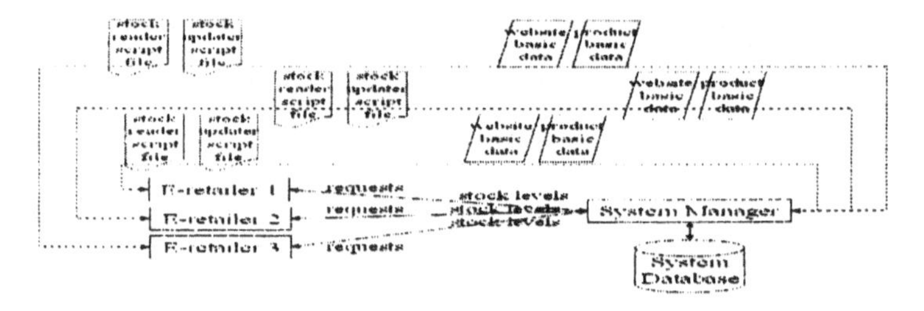

Figure 1: The operation procedure of the OCSM mechanism.

```
<!--#INCLUDE FILE="[illegible]"-->
<%
DIM N,TITLES,TITLESNEW,STOCKS,ORIGINALSTOCKS
N=2
TITLES="'TravelMate [illegible]','TravelMate [illegible]'"
TITLESNEW=""
STOCKS=""
ORIGINALSTOCKS=""
SET RS=SERVER.CREATEOBJECT("ADODB.RECORDSET")
RS.OPEN "SELECT name,amount,stock FROM product WHERE name IN ("&TITLES&")",CONN,1,3
DO WHILE NOT RS.EOF
TITLESNEW=TITLESNEW+SPACE(20-LEN(RS("name")))+RS("name")
STOCKS=STOCKS+SPACE(10-LEN(CSTR(RS("stock"))))+CSTR(RS("stock"))
ORIGINALSTOCKS=ORIGINALSTOCKS+SPACE(10-LEN(CSTR(RS("amount"))))+CSTR(RS("amount"))
RS.MOVENEXT
LOOP
RS.CLOSE
SET RS=NOTHING
CONN.CLOSE
SET CONN=NOTHING
RESPONSE.REDIRECT "HTTP://[illegible]/OCSM/STOCK_RECORDER.ASP?
WEBTITLE=TolyShop&N="&N&"&TITLESNEW="&TITLESNEW&"&STOCKS="&STOCKS&"&ORIGINALSTOCKS="&ORIGINALSTOCKS
%>
```

Figure 2: The script file of a stock reader.

3. An Experimental System for Demonstration

To demonstrate the applicability of the proposed OCSM mechanism, an experimental system composed of two e-retailers and a system server has been constructed. Both e-retailers were constructed using the TimeShop shopping system (refers to http://www.vipasp.com/)(see Figure 3). The platform of the system server is Windows XP and Microsoft Internet Information Server v. 5.1. The system database management system is Microsoft Access XP with ODBC. The system manager is composed of active server pages (ASP) developed with VBScript programming language.

Figure 3: An e-retailer in the experimental system.

The operation procedure is as follows. Firstly an e-retailer goes to the system register to register its basic data (see Figure 4). After checking the validity of the registration data, the e-retailer is informed of a set of username and password to register new products that will be sold with the mechanism (also see Figure 4). The e-retailer can also join the collaborative sale of other products that have been involved in the mechanism (see Figure 5). Every a small time interval, the request sender of the system manager sends a request to run the stock reader on each participating website, and the stock reader will respond the current stock levels of products involved in the OCSM mechanism on the website. Based on these data, the total stocks levels of these products can be obtained. After that, the request sender of the system manager sends a request to run the stock updater on each participating website to update the stock levels.

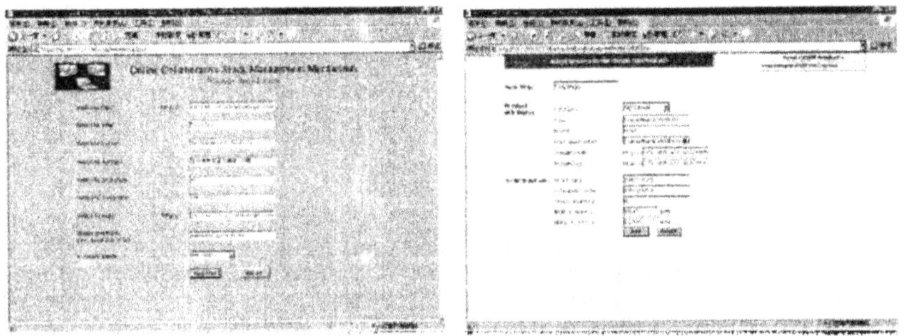

Figure 4: E-retailer registration (left-hand side) and product registration (right-hand side).

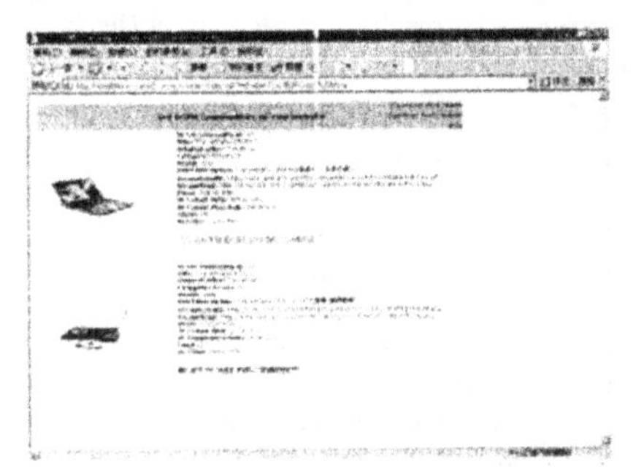

Figure 5: An e-retailer can join the collaborative sale of other products.

4. Conclusions

An OCSM mechanism for multiple e-retailers is proposed, in which the stocks on these websites are monitored, virtually aggregated, and shared among websites. As a result, it becomes possible for an e-retailer to sell more than the stock, and the sales are increased. On the other hand, the OCSM mechanism provides all participating websites with more online sale channels, and the possibilities of selling products are therefore improved. An experimental system has also been constructed to illustrate the practicability of the OCSM mechanism.

References

[1] V. Zwass, Electronic commerce and organizational innovation: aspects and opportunities, *International Journal of Electronic Commerce* 7 (3) 7-37 (2003).

[2] R. T. Wigand, Electronic commerce definition, theory and context, *The Information Society* 13 (1) 1-16 (1997).

[3] T. Ito, H. Hattori, T. Shintani, A cooperative exchanging mechanism among seller agents for group-based sales, *Electronic Commerce Research and Applications* 1 138-149 (2002).

Brill Academic Publishers
P.O. Box 9000, 2300 PA Leiden,
The Netherlands

Lecture Series on Computer and Computational Sciences
Volume 2, 2005, pp. 24-27

Symbolic Semantics for Weak Open Bisimulation in π-calculus[1]

Taolue Chen[2] Tingting Han Jian Lu

State Key Laboratory of Novel Software Technology,
Department of Computer Science and Technology
Nanjing University, Nanjing, Jiangsu, P.R.China 210093

Received 25 March, 2005; accepted in revised form 7 April, 2005

Abstract: This paper investigates the symbolic semantics of the weak open congruence on finite π processes with the mismatch operator. The standard definition of weak open congruence gives rise to an ill-defined equivalence in the presence of the mismatch operator. Two alternatives, that is, the symbolic weak late open congruence and symbolic weak early open congruence are proposed. We show that they coincide with the non-symbolic version late and early open congruence respectively. Axiomatization systems for the two weak open congruences are given, which are essentially the symbolic proof system for strong open congruence with the different tau laws. The soundness and completeness of these proof systems are shown. The results can be seen as the basis of automated algorithms and tools to check the weak open bisimulation and give more insights into the symbolic semantics and bisimilation theory.

Keywords: Process Algebra, Mobile Process, Open Bisimulation, Symbolic Semantics, Axiomatization.

1 Introduction

Over the last decades, various calculi of mobile processes, notably the π-calculus [4], have been the focus of research in concurrency theory. In the study of algebra theory for mobile process, many researchers focus on the bisimulation equivalence between processes. For π-calculus people have introduced many kinds of them. such as late and early congruences [4]. Among these, the open bisimulation [6] is one of the most significant and interesting. Generally speaking, there are two main approaches for the semantics of mobile processes, i.e. the symbolic approach [2] and the non-symbolic approach [5]. In this paper, we focus on the former. To the authors' knowledge, hitherto, the symbolic semantics and axiomatization system for *weak open congruence* in the presence of the *mismatch* operator have not been studied in depth. The main work of this paper aims to close this open problem. Recently, in the non-symbolic framework, such problem has been investigated by Fu and Yang. see [1]. There are at least two reasons for studying symbolic semantics for π-calculus. On the one hand, symbolic framework often can yield alternative, more efficient characterizations and complete proof systems to reason about them; on the other hand, symbolic technique will shed new lights on the differences between all sorts of bisimulations on the basis of diverse instantiation strategies. In a word, to provide the definition and proof system can definitely convey us more insights into the algebra theory of π-calculus with mismatch, especially the theory of bisimulation.

[1]Supported by 973 Program of China (2002CB312002) and NNSFC (60233010, 60273034, 60403014), 863 Program (2002AA116010).

[2]Corresponding author. E-mail: ctl@ics.nju.edu.cn

Now, let us turn to the technical aspect. As pointed out by Fu and Yang, the standard definition of weak open congruence gives rise to an ill-defined equivalence in the presence of the mismatch operator. Consider the following counterexample. Let $P = a(x).[x \neq y]\tau.R + a(x).R$ and $Q = a(x).[x \neq y]\tau.R$, it is not difficult to check that P and Q are "weak open congruent" when standard definition is adopted. However, $(a(x).[x \neq y]\tau.R + a(x).R) \mid \bar{a}y$ and $(Q = a(x).[x \neq y]\tau.R) \mid \bar{a}y$ are not congruent, hence the definition is not sound in the sense that for a bisimulation relation, closing by parallel operator is a basic requirement. The anomaly is mainly caused by the delay of instantiation of input actions, which would not have been a great problem in the absence of the mismatch operator. So, with the mismatch operator around it seems better to let the instantiation happen immediately. From the symbolic point of view, we must refine the instantiation strategy of open bisimulation, that is, subsume some strategies of late and early bisimulations. This should be reflected by the restriction imposed on the case partition. Based on the above ideas, the *symbolic late open congruence* and *symbolic early open congruence* are proposed as two alternatives. We show that they coincide with the non-symbolic version late open congruence and early open congruence in [1] respectively. Axiomatization systems for the two weak open congruences are also given in the symbolic framework. They are essentially the symbolic proof system for strong open congruence with the different tau laws. The soundness and completeness are shown. Note that in this extended abstract, due to space restriction, most preliminary materials on π-calculus are omitted, we refer the people to [4, 3] for more details. Moreover, all the proofs are omitted.

2 Symbolic Weak Late Open Congruence

In this section, we give the definition for the *symbolic weak late open congruence*. We will show that such a definition coincides with the concrete version due to Fu and Yang. In the sequel, we denote *concrete weak late transition relations* as $\stackrel{\alpha}{\Rightarrow}_l$, *symbolic weak late transition relations* as $\stackrel{\phi,\alpha}{\Rightarrow}_L$. The definition for $\alpha =^{\phi} \beta$ is also standard. For details, we refer the readers to [3].

Definition 1 *A condition indexed family of symmetric relations* $\mathcal{S} = \{S^{\phi}\}$ *is a* symbolic weak late open bisimulation *if* $(P, Q) \in S^{\phi}$ *implies:*

whenever $P \stackrel{\psi,\alpha}{\longrightarrow} P'$, *then for each* $\phi' \in MCE_{fn(P,Q)}(\psi)$, *if* $\phi \wedge \phi'$ *is consistent, then there is a* $Q \stackrel{\psi',\hat{\beta}}{\Rightarrow}_L Q'$ *such that* $\phi \wedge \phi' \Rightarrow \psi'$, $\alpha =^{\phi\wedge\phi'^{+}} \beta$ *and*

- *If* $\alpha \equiv a(x)$, *then for each* $\phi'' \in MCE_{fn(P,Q)\cup\{x\}}(\phi')$, *if* $\phi \wedge \phi''$ *is consistent, then there is a* $Q' \stackrel{\psi'',\varepsilon}{\Rightarrow}_L Q''$. *s.t.* $\phi \wedge \phi'' \Rightarrow \psi''$ *and* $(P', Q'') \in S^{\phi\wedge\phi''^{+}}$.
- *If* $\alpha \equiv \bar{a}(x)$, *then* $(P', Q') \in S^{\phi\wedge\phi'^{+}\wedge\bigwedge\{x\neq y \mid y\in fn(P,Q)\}}$.
- *If* α *is a free action, then* $(P', Q') \in S^{\phi\wedge\phi'^{+}}$.

P *is* weak late open bisimular *to* Q *w.r.t.* ϕ, *notation* $P \approx^{\phi} Q$ *if there exists a* symbolic weak late bisimulation $\mathcal{S} = \{S^{\phi}\}$ *such that* $(P, Q) \in S^{\phi}$. *We say that* P *is* symbolic weak late open bisimular *to* Q *if* $P \approx^{\emptyset} Q$.

The following theorem states the coincidence of concrete and symbolic bisimulation relations. The definition for concrete version can be found in [1]. Note that for convenience, for distinction D [4], condition ϕ [3], we denote D^{-} as $\bigwedge\{x \neq y \mid (x, y) \in D\}$, ϕ^{-} as $\bigwedge\{x \neq y \mid \phi \Rightarrow x \neq y\}$ and ϕ^{+} as $\bigwedge\{x = y \mid \phi \Rightarrow x = y\}$.

Theorem 2 $P \approx^{\phi} Q$ *iff* $P\sigma \approx^{D} Q\sigma$ *where* $D = \phi^{-}\sigma$.

EQUIV $\dfrac{-}{true \rhd P = Q} \quad \dfrac{true \rhd P = Q}{true \rhd Q = P} \quad \dfrac{true \rhd P = Q \quad true \rhd Q = R}{true \rhd P = R}$

AXIOM $\dfrac{-}{true \rhd P = Q}$ P, Q are the instantiation of axiom schemas

MATCH $\dfrac{\phi \wedge (x = y) \rhd P = Q \quad \phi \wedge (x \neq y) \rhd 0 = Q}{\phi \rhd [x = y]P = Q}$ TAU $\dfrac{\phi \rhd P = Q}{\phi \rhd \tau.P = \tau.Q}$

L_INPUT $\dfrac{\phi \rhd P = Q}{\phi \rhd a(x).P = b(x).Q}$ $\phi \Rightarrow a = b \wedge x \notin n(\phi)$

OUTPUT $\dfrac{\phi \rhd P = Q \quad \phi \Rightarrow (a = b) \wedge (x = y)}{\phi \rhd \bar{a}x.P = \bar{b}y.Q}$ CONSEQ $\dfrac{\phi \rhd P = Q}{\psi \rhd P = Q}$ $\psi \Rightarrow \phi$

CHOICE $\dfrac{\phi \rhd P_i = Q_i \quad i = 1,2}{true \rhd P_1 + P_2 = Q_1 + Q_2}$ ABSURD $\dfrac{-}{false \rhd P = Q}$

CONDITION $\dfrac{\phi \rhd P = Q}{\phi \rhd \psi_1 P = \psi_2 Q}$ $\phi \wedge \psi_1 \Leftrightarrow \phi \wedge \psi_2$ RES $\dfrac{\phi \rhd P = Q}{\nu_x \phi \rhd (\nu x)P = (\nu x)Q}$

Figure 1: Inference System for Weak Late Open Congruence.

S1	$X + 0 = X$	S2	$X + X = X$
S3	$X + Y = Y + X$	S4	$(X + Y) + Z = X + (Y + Z)$
R1	$(\nu x)0 = 0$	R2	$(\nu x)\alpha.X = \alpha.(\nu x)X \quad x \notin n(\alpha)$
R3	$(\nu x)\alpha.X = 0 \quad \text{if } x = \text{subj}(\alpha)$	R4	$(\nu x)(\nu y)X = (\nu y)(\nu x)X$
R5	$(\nu x)(X + Y) = (\nu x)X + (\nu x)Y$	MS1	$X = [x = y]X + [x \neq y]X$
MS2	$[x \neq x]X = 0$	MS3	$(\nu x)[y \neq z]X = [y \neq z](\nu x)X$
MS4	$[x \neq y](X + Y) = [x \neq y]X + [x \neq y]Y$		

Figure 2: Axiom Schema for Weak Late Open Congruence.

It is clear that $\approx^{\phi}$ is not a congruence relation. The canonical way to obtain a congruence from bisimulation equivalence is to take the largest congruence contained in it. See Definition 3.

Definition 3 *Two processes P and Q are* symbolic weak late open ϕ-congruent, *notation $P \simeq^{\phi} Q$, if $P \approx^{\phi} Q$ and the following conditions are satisfied:*

- *$P \overset{\psi,\tau}{\rightarrow} P'$, then for each $\phi' \in MCE_{fn(P,Q)}(\psi)$, if $\phi \wedge \phi'$ is consistent, then there is a $Q \overset{\psi',\tau}{\Rightarrow}_L Q'$ such that $\phi \wedge \phi' \Rightarrow \psi'$ and $P' \approx^{\phi \wedge \phi'^{+}} Q'$.*

- *$Q \overset{\psi,\tau}{\rightarrow} Q'$, then for each $\phi' \in MCE_{fn(P,Q)}(\psi)$, if $\phi \wedge \phi'$ is consistent, then there is a $P \overset{\psi',\tau}{\Rightarrow}_L P'$ such that $\phi \wedge \phi' \Rightarrow \psi'$ and $Q' \approx^{\phi \wedge \phi'^{+}} P'$.*

3 Proof System

This section is devoted to formulating a proof system for *symbolic weak late open congruence* and proving its soundness and completeness. We let AS denote the system consisting of the well-known *Expansion Law* [4] and the inference system reported in Figure 1, the axiom schema reported in Figure 2 with the tau laws reported in Figure 3. We write $AS \vdash \phi \rhd P = Q$ to mean that $\phi \rhd P = Q$ can be derived from AS. The soundness of the system is obvious. For the completeness, we have:

T1	$\alpha.\tau.P$	$=$	$\alpha.P$	
T2	$P+\tau.P$	$=$	$\tau.P$	
T3	$\alpha.(P+\psi\tau.Q)$	$=$	$\alpha.(P+\psi\tau.Q)+\psi\alpha.Q$	$bn(\alpha)\cap n(\psi)=\emptyset$
T4	$\tau.P$	$=$	$\tau.(P+\psi\tau.P)$	

Figure 3: Tau Laws for Weak Late Open Congruence.

4 Early Case

In this section, we turn to the early case. The situation is similar to that in the late scenario.

Definition 4 *A condition indexed family of symmetric relations* $\mathcal{S} = \{S^\phi\}$ *is a* symbolic weak early open bisimulation *if* $(P,Q) \in S^\phi$ *implies:*

whenever $P \overset{\psi,\alpha}{\rightarrow} P'$*, then for each* $\phi' \in MCE_V(\psi)$*, if* $\phi \wedge \phi'$ *is consistent, then there is a* $Q \overset{\psi',\beta}{\Rightarrow} Q'$ *such that* $\phi \wedge \phi' \Rightarrow \psi'$, $\alpha =^{\phi\wedge\phi'^+} \beta$ *and*

- *If* $\alpha \equiv a(x)$*, then* $V = fn(P,Q) \cup \{x\}$ *and* $(P',Q') \in S^{\phi\wedge\phi'^+}$.
- *If* $\alpha \equiv \bar{a}(x)$*, then* $V = fn(P,Q)$ *and* $(P',Q') \in S^{\phi\wedge\phi'^+\wedge\bigwedge\{x\neq y|y\in fn(P,Q)\}}$.
- *If* α *is a free action, then* $V = fn(P,Q)$ *and* $(P',Q') \in S^{\phi\wedge\phi'^+}$.

P *is* symbolic weak early open bisimular *to* Q *w.r.t.* ϕ*, notation* $P \approx_E^\phi Q$ *if there exists a* symbolic weak early open bisimulation $\mathcal{S} = \{S^\phi\}$ *such that* $(P,Q) \in S^\phi$*. We say that* P *is* symbolic weak early open bisimular *to* Q *if* $P \approx_E^\emptyset Q$.

The notion of *symbolic weak early open congruent*, denoted by $\simeq_E^\phi$ can be defined in the similar way as in the Section 2. It can also be shown that it coincides with the corresponding concrete version.

Proof System. The proof system for *symbolic weak early open congruence* can be obtained by adding the following T5 law to the proof system AS.

$$\text{T5} \quad \sum_{i\in I} a(x).(P_i + \psi_i\tau.Q) = \sum_{i\in I} a(x).(P_i + \psi_i\tau.Q) + \psi a(x).Q \qquad \bigvee_{i\in I} \psi_i \Leftrightarrow \psi, x \notin n(\psi)$$

We use AS_w to denote $AS \cup \{\text{T5}\}$. The soundness of AS_w is clearly. And we have:

Theorem 5 *(Completeness of* AS_w*) If* $P \simeq_E^\phi Q$*, then* $AS_w \vdash \phi \triangleright P = Q$.

References

[1] Y.Fu, Z.Yang. Tau Laws for Pi Calculus. Theor. Comput. Sci, 308(1-3): 55-130, 2003.

[2] M.Hennessy, H.Lin. Symbolic bisimulations. Theor. Comput. Sci, 138(2): 353-389, 1995.

[3] H.Lin. Complete inference systems for weak bisimulation equivalences in the π-calculus, Inf. Comput, 180(1): 1-29, 2003.

[4] R.Milner, J.Parrow, D.Walker. A Calculus of Mobile Process, part I/II. Journal of Information and Computation, 100:1-77, 1992.

[5] J.Parrow, D.Sangiorgi. Algebraic theories for name-passing calculi. Inf. Comput, 120(2): 174-197,1995.

[6] D.Sangiorgi. A theory of bisimulation for π-calculus, Acta Informatica 33(1): 69-97, 1996.

Brill Academic Publishers
P.O. Box 9000, 2300 PA Leiden,
The Netherlands

Lecture Series on Computer and Computational Sciences
Volume 2, 2005, pp. 28-31

A Fuzzy Set Approach for Analyzing Customer RFM Data

T. Chen Hsin-Chieh Wu[2]

[1]Department of Industrial Engineering and System Management,
Feng Chia University,
Taichung, Taiwan 407, ROC
[2]Department of Industrial Engineering and Management,
Chaoyang University of Technology
Taichung County, Taiwan 413, ROC

Received 27 March, 2005; accepted in revised form 7 April, 2005

Abstract: The RFM model is an important method in customer importance evaluation. However, there are some shortcomings in traditional crisp RFM models. To overcome these shortcomings, a fuzzy set approach is proposed. Firstly, the formula for calculating the R, F, and M scores as well as that for calculating the RFM total score are fuzzified, and a customer's importance is now evaluated with a more reasonable fuzzy value instead. Secondly, a formula for calculating the fuzzy weighted RFM total score is also proposed to deal with the situation in which R, F, and M are considered unequally important. After screening out unimportant customers with low fuzzy (weighted) RFM total scores, the Fuzzy C-means approach is applied to cluster important customers, so as to establish different marketing strategies. A demonstrative example is given. Advantages of the fuzzy set approach over traditional methods are then concluded.

Keywords: RFM model; Customer importance; Fuzzy C-means; Fuzzy set approach; CRM

1. Introduction

Customer clustering and customer importance evaluation are two important topics in customer relationship management (CRM). In these fields, the RFM model is a very famous approach. The R in RFM is the recency of the latest purchase; F is the purchasing frequency; and M is the total monetary amount. The RFM model evaluates a customer's importance with the RFM total score, and classifies customers according to the different buying power and magnitude of each customer cluster. Traditional RFM models include the models of Stone [3], Miglautsch [2], Sung and Sang [4], etc. The shortcomings of traditional RFM models are:

(1) The way of dividing a dimension (R, F, or M) into equal sections might not be suitable for all products.
(2) Scoring with crisp numbers is too subjective and inflexible.
(3) The exploration of the content of each customer cluster provides little information.
(4) It is difficult to differentiate customers with the same RFM total scores.

To overcome these shortcomings, a fuzzy set approach is proposed in this paper. Firstly, the formula for calculating the R, F, and M scores as well as that for calculating the RFM total score are fuzzified, and a customer's importance is now evaluated with a more reasonable fuzzy value instead. A fuzzy RFM total score not only preserves the subjective and imprecise nature of a customer's "importance", but also provides some information to assist the determination of marketing targets. Secondly, a formula for calculating the fuzzy weighted RFM total score is also proposed to deal with the situation in which R, F, and M are considered unequally important. After screening out unimportant customers with low fuzzy (weighted) RFM total scores, the Fuzzy C-means (FCM) approach is applied to cluster important customers, so as to establish different marketing strategies. In this way, the number of

[1] Corresponding author E-mail: tolychen@ms37.hinet.net

customer clusters can be arbitrarily specified, considering the scarcity of marketing resources and the diversification of marketing strategies. At last, a demonstrative example is given, and the advantages of the proposed methodology over traditional RFM models are then concluded.

2. The Fuzzy R, F, M scores and the Fuzzy RFM Total Score

Calculating each customer's RFM total score can help to find out the most important customers to the enterprise as the targets of marketing. However, a customer's importance is a subjective and imprecise concept, and it will be more appropriate if the importance can be expressed with a fuzzy value. The following procedure is proposed to calculate the fuzzy RFM total score of a customer:

(1) Apply simple fuzzy partition to divide each dimension, and then assign a score to each fuzzy interval: The data range from the minimum to the maximum is divided into fuzzy intervals with the same widths.
(2) Calculate the membership that the customer's R, F, or M value belongs to each fuzzy interval, so as to obtain the customer's fuzzy score, which can be expressed as {(score, membership)}.
(3) Add up the customer's fuzzy R, F, and M scores according to Zadeh's extension principle:

$$\tilde{T} = \tilde{R}(+)\tilde{F}(+)\tilde{M} \tag{1}$$

$$\mu_{\tilde{T}}(t) = \sup_{t=r+f+m} \min(\mu_{\tilde{R}}(r), \mu_{\tilde{F}}(f), \mu_{\tilde{M}}(m)) \tag{2}$$

The fuzzy RFM total score $\tilde{T}$ can be defuzzified according to Wrather and Yu's formula if necessary:

$$D(\tilde{T}) = \sum_{all\ t} t\mu_{\tilde{T}}(t) / \sum_{all\ t} \mu_{\tilde{T}}(t) \tag{3}$$

3. The Fuzzy Weighted RFM Total Score

A formula for calculating the fuzzy weighted RFM total score is proposed to deal with the situation in which R, F, and M are considered unequally important. The procedure is as follows:

(1) Adopt linguistic values to define fuzzy weights: For example, {"Very Unimportant", "Unimportant", "Moderate", "Important", "Very important"} = {(0, 0, 3), (0, 0.3, 0.5), (0.2, 0.5, 0.8), (0.5, 0.7, 1), (0.7, 1, 1)} in the form of triangular fuzzy numbers (TFNs).
(2) Specify the fuzzy weights for the three dimensions (R, F, and M).
(3) Decompose the membership functions of the three fuzzy weights into combinations of values and memberships.
(4) Multiply three times the fuzzy weights to the customer's fuzzy scores, and then divide the result by the fuzzy sum of the fuzzy weights:

$$\tilde{T}_W = 3 \cdot (\tilde{W}_R(\times)\tilde{R}(+)\tilde{W}_F(\times)\tilde{F}(+)\tilde{W}_M(\times)\tilde{M}) / (\tilde{W}_R(+)\tilde{W}_F(+)\tilde{W}_M) \tag{4}$$

$$\mu_{\tilde{T}_w}(t_w) = \sup_{t_w = w_R r + w_F f + w_M m} \min(\mu_{\tilde{R}}(r), \mu_{\tilde{F}}(f), \mu_{\tilde{M}}(m), \mu_{\tilde{W}_R}(w_R), \mu_{\tilde{W}_F}(w_F), \mu_{\tilde{W}_M}(w_M)) \tag{5}$$

The fuzzy weighted RFM total score can also be defuzzified if necessary.

4. Applying the FCM Approach to Cluster Important Customers

The FCM approach performs clustering by minimizing the following objective function:

$$\text{Min} \sum_{k=1}^{K} \sum_{i=1}^{n} \mu_{i(k)}^{m} e_{i(k)}^{2} \tag{6}$$

where K is the required number of clusters; n is the number of customers; $\mu_{i(k)}$ represents the membership of customer i belonging to cluster k; $e_{i(k)}$ measures the distance from customer i to the

centroid of cluster k; $m \in (1, \infty)$ is a parameter to increase or decrease the fuzziness. The procedure of applying the FCM approach to cluster important customers is as follows:
(1) Screen out unimportant customers with low fuzzy (weighted) RFM total scores.
(2) Establish an initial clustering result.
(3) (Iterations) Obtain the centroid of each cluster as:

$$\bar{x}_{(k)} = \{\bar{x}_{(k)j}\} \tag{7}$$

$$\bar{x}_{(k)j} = \sum_{i=1}^{n} \mu_{i(k)}^{m} x_{ij} / \sum_{i=1}^{n} \mu_{i(k)}^{m} \tag{8}$$

$$\mu_{i(k)} = 1 / \sum_{l=1}^{K} (e_{i(k)} / e_{i(l)})^{2/(m-1)} \tag{9}$$

$$e_{i(k)} = \sqrt{\sum_{all\ j} (x_{ij} - \bar{x}_{(k)j})^2} \tag{10}$$

where $\bar{x}_{(k)}$ is the centroid of cluster k.

(4) Re-measure the distance of each customer to the centroid of every cluster, and then recalculate the corresponding membership.
(5) Stop if the following condition is satisfied. Otherwise, return to step (2):

$$\max_{k} \max_{i} | \mu_{i(k)}^{(t)} - \mu_{i(k)}^{(t-1)} | < d \tag{11}$$

where $\mu_{i(k)}^{(t)}$ is the membership of customer i belonging to cluster k after the t-th iteration; d is a real number representing the threshold of membership convergence. Finally, the separate distance test (S test) proposed by Xie & Beni [5] can be applied to determine the optimal number of clusters K.

5. A Demonstrative Example

A sample database of Microsoft Access 2000, the Northwind Traders database, is adopted to demonstrate the application of the proposed fuzzy set approach. There are 89 customers in this example. Firstly, the fuzzy RFM total score of each customer is calculated to evaluate the customer's importance. The results are shown in Table 1. The fuzzy RFM total score also allows of some flexibility in determining the targets of marketing. For example, if the marketing resources are very sufficient, then even customers whose memberships of high RFM total scores are greater than 0 can be included. Otherwise, customers whose memberships of low RFM total scores are greater than 0 should be excluded. The fuzzy RFM total score can also be defuzzified if necessary.

Table1: The fuzzy RFM total scores of the customers.

Customer	Fuzzy RFM total score {(value, membership)}	Defuzzified result
SUPRD	{(11, 0.11), (12, 0.6), (13, 0.4)}	12.3
ERNSH	{(9, 0.01), (10, 0.02), (11, 0.49), (12, 0.51)}	11.5
SAVEA	{(10, 0.04), (11, 0.61), (12, 0.39)}	11.3
...	...	...

Secondly, assume R, F, and M are assigned different weights: R is "Moderate", F is "Important", and M is "Very Important". Then the fuzzy weighted RFM total score of each customer can be calculated to evaluate the customer's importance. The results are summarized in Table 2. The top 3 most important customers in Table 1 and Table 2 are the same. Besides, the spread of the fuzzy weighted RFM total score is much wider than that of the fuzzy RFM total score, which indicates that more flexibility is gained by assigning different weights.

Table 2: The fuzzy weighted RFM total scores of the customers.

Customer	Fuzzy weighted RFM total score	Defuzzified result
SUPRD	{(7.6, 0.11), ..., (19.5, 0.33)}	13.1
ERNSH	{(5.1, 0.01), ..., (17.6, 0.33)}	11.3
SAVEA	{(6, 0.04), ..., (17.6, 0.33)}	11.1
...	...	...

Thirdly, the FCM approach is applied to cluster important customers that are chosen with two criteria: the defuzzified result of the fuzzy weighted RFM total score is greater than 8; the membership of high fuzzy weighted RFM total scores (e.g. > 14) is greater than 0. As a result, there are only 26 important customers. After normalizing their RFM data, the FCM approach is applied to classify them into 2~6 clusters. According to the result of the S test, the optimal number of customer clusters is $K = 4$, and every important customer can be classified into the four clusters with different memberships, which allows of the flexibility in determining the targets of marketing. If we classify every customer into the cluster with the highest membership absolutely, then the clustering result is as shown in Figure 1. After exploring the contents of the four clusters, different marketing strategies can be established.

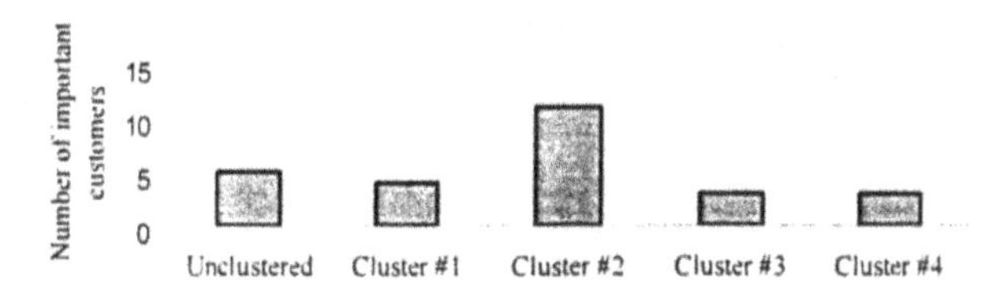

Figure 1: The clustering result of four important customer clusters.

6. Conclusions

To deal with the problems in traditional crisp RFM models, a fuzzy set approach is proposed in this paper. Firstly, the formula for calculating the R, F, and M scores as well as that for calculating the RFM total score are fuzzified. Secondly, a formula for calculating the fuzzy weighted RFM total score is also proposed to deal with the situation in which R, F, and M are considered unequally important. After screening out unimportant customers with low fuzzy (weighted) RFM total scores, the Fuzzy C-means approach is applied to cluster important customers. The conclusions of this paper are:

(1) Evaluating a customer's importance with the fuzzy-valued RFM total score provides some information to assist the determination of marketing targets, and elevates the flexibility of the task.

(2) Traditional ways of classifying customers are too subjective and inflexible. Conversely, in the FCM approach, a customer can be simultaneously classified into more than one cluster with different memberships, which increases the flexibility in choosing the targets for marketing.

(3) More flexibility can be gained by assigning different weights to R, F, and M.

References

[1] H. P. Chiu and C. Y. Su, A flexible fuzzy RFM model for CRM and database marketing, *Journal of Electronic Business* 6 (2) 1-26 (2004).

[2] J. Miglautsch, Thoughts on RFM scoring, *Journal of Database Marketing* 8 (1) (2000).

[3] B. Stone: *Successful Direct Marketing Methods*, NTC Business Books, 1989.

[4] H. H. Sung and C. P. Sang, Application of data mining tools to hotel data mart on the Intranet for database marketing, *Expert Systems with Applications* 15 1-31 (1989).

[5] X. L. Xie and G. Beni, A validity measure for fuzzy clustering, *IEEE Transactions of Pattern Analysis and Machine Intelligence* 13 (1991).

Brill Academic Publishers
P.O. Box 9000, 2300 PA Leiden,
The Netherlands

Lecture Series on Computer and Computational Sciences
Volume 2, 2005, pp. 32-36

Towards A Bisimulation Congruence for ROAM [1]

Taolue Chen [a,b, 2] **Tingting Han** [a] **Jian Lu** [a]

[a] State Key Laboratory of Novel Software Technology,
Department of Computer Science and Technology
Nanjing University, Nanjing, Jiangsu, P.R.China 210093
[b] CWI, Department of Software Engineering,
PO Box 94079, 1090 GB Amsterdam, The Netherlands

Received 10 April, 2005; accepted in revised form 25 April, 2005

Abstract: Computation with mobility becomes a novel distributed computation paradigm with the development of network technology. The calculus of *Mobile Ambient* (MA) is a widely studied formal mechanism for describing both mobile computing and mobile computation. This paper deals with the algebra theory of a variant of this calculus, the *Robust Mobile Ambient* (ROAM). Hitherto, for ROAM, the criterion of process equivalence is contextual equivalence, whose main disadvantage lies in that the proof and verification for process equivalence are usually very hard. To remedy this deficiency to some extend, we provide a new labelled transition system, and based on it, a bisimulation relation is introduced as the criterion of behavior equivalence. We justify the soundness of definition by showing that it is a congruence relation and to illustrate the advantage of the new behavior equivalence relation, this paper also gives concise proofs for some important process equivalence laws studied in literature.

Keywords: Process algebra, Mobile ambient, Label transition system, Bisimulation

1 Introduction

Computation with mobility is a kind of new distributed computation paradigm with the development of network technology. The Calculus of *Mobile Ambient* ([1], MA in short), introduced by Cardelli and Gordon, is a formalism which can describe both of the two aspects of mobility within a single framework, and is one of the hottest topics in the area of mobile process calculi after π-calculus [3]. However, in [4], Levi and Sangiorgi have pointed out that there exists a kind of dangerous interference in the original ambient calculus, which shows some deficiency of original design for MA. To settle this problem, a new calculus called *Mobile Safe Ambient* ([4], SA in short) is presented in order to eliminate this kind of interferences through adding the corresponding co-action primitives to the original action primitives of MA. However, as pointed out by Guan et al [2], the co-actions introduced in SA are very vulnerable to tampering in that some malicious third parties may easily consume them. Thus a variant of MA, called *Robust Ambient* (ROAM in short), is introduced, which can explicitly name the ambient that can participate in the reduction and eliminate grave interference problem, too. At the same time, it allows a finer grained and more natural access control.

[1]Supported by 973 Program of China (2002CB312002) and NNSFC (60233010, 60273034, 60403014), 863 Program (2002AA116010).

[2]Corresponding author. E-mail: ctl@ics.nju.edu.cn

For mobile ambient calculus and its variation, the current research mainly focus on two aspects. One is the equation theory for calculus, e.g [1], the main topic is how to define and prove the equivalence between process. The other is the type system of calculus, e.g. [5], the main topic is to introduce appropriate type system according to different requirements. This paper mainly deals with the first problem. We refer the reader to [2] for more details on ROAM. It is worth pointing out that in [2], the definition on equivalence is so called contextual equivalence, correspondingly, the labelled transition system used there is defined based on the harden relation. In fact, this is a criterion based on testing. Note that in the research for common process algebra, e.g. π-calculus, the equivalence definition based on bisimulation is more general, whose advantages lie in that it has clearer intuition sense and some co-inductive proof techniques, e.g. *up to technique* can be used, thus we can give neater proof for some important process equivalence laws and provide more efficient model checking algorithms. Base on the above ideas, this paper deals with the behavior equivalence from the point view of bisimulation. In detail, at first, we provide a new labelled transition system by introducing concretion, and the coincidence between the reduction semantics and the labelled transition semantics is proved; then a process equivalence definition based on bisimulation is given and we prove that it is a congruence relation, which justifies the soundness of the definition in some sense. In order to show the usage of our definition, we also give neat proofs of some key algebra laws for process equivalence.

Note that in this extended abstract, due to space restriction, most preliminary materials on ROAM are omitted, we refer the people to [4, 2] for more details. Moreover, most of the proofs are omitted. Here, we only give a very brief of syntax of ROAM, which is recursively defined as follows:

$$P ::= 0 \mid (\nu n)P \mid P|Q \mid M.P \mid n[P] \mid !P$$

where, M is element of capability sets (denoted by CAP) which is recursively defined as:

$$M ::= in\langle n\rangle \mid out\langle n\rangle \mid open\langle n\rangle \mid \overline{in}\langle n\rangle \mid \overline{out}\langle n\rangle \mid \overline{open}$$

where, $n \in \mathcal{N}$.

2 Labelled Transition System

In this section, we provide a labelled transition system by introducing concretion, which is essential for the definition of process equivalence. Comparing to the traditional process calculus, such as π-calculus, it is much more difficult to define similar labelled transition system. The key point lies in that in order to describe the whole behavior of process, not only the original prefix in the syntax of calculus, such as *in*, *out open*, but also the behavior of the ambient which contains these actions must be considered. Thus we have to add some auxiliary actions to describe the behavior of ambient. However, it will make the system much more intricate since a process, say P, after doing some action α, can convert not only to a process, but also to a concretion, which is in the form of $\nu\tilde{m}\langle P\rangle_n Q$. In intuition, process P is the residual section and n the target, while Q is the migrating ambient and $\tilde{m}$ is a group of names.

As in the common labelled transition system, first we define the prefixes, actions and concretion by BNF as follows:

Prefix:	$\mu ::= in\langle n\rangle \mid out\langle n\rangle \mid open\langle n\rangle \mid \overline{in}\langle n\rangle \mid \overline{out}\langle n\rangle \mid \overline{open}$
Action:	$\alpha ::= \tau \mid \mu \mid enter\langle n,m\rangle \mid \overline{enter}\langle n,m\rangle \mid exit\langle n,m\rangle \mid ?n$
Concretion:	$K ::= \nu\tilde{m}\langle P\rangle_n Q$
Outcome:	$O ::= P \mid K$

The labelled transition system is defined in Table 1. The following theorem shows the coin-

$$
\begin{array}{llll}
\text{(Act)} & \dfrac{}{\mu.P \xrightarrow{\mu} P} & \text{(Repl act)} & \dfrac{}{!\mu.P \xrightarrow{\mu} P|!\mu.P} \\
\text{(Enter)} & \dfrac{P \xrightarrow{in\langle n\rangle} P'}{m[P] \xrightarrow{enter\langle n,m\rangle} \langle m[P']\rangle_n 0} & \text{(Co-enter)} & \dfrac{P \xrightarrow{\overline{in}\langle m\rangle} P'}{n[P] \xrightarrow{\overline{enter}\langle n,m\rangle} \langle P'\rangle_n 0} \\
\text{(Exit)} & \dfrac{P \xrightarrow{out\langle n\rangle} P'}{m[P] \xrightarrow{exit\langle n,m\rangle} \langle 0\rangle_n m[P']} & \text{(?-out)} & \dfrac{P \xrightarrow{exit\langle n,m\rangle} \nu\tilde{p}\langle P_1\rangle_n P_2 \quad Q \xrightarrow{\overline{out}\langle m\rangle} Q'}{P|Q \xrightarrow{?n} \nu\tilde{p}\langle P_1|Q'\rangle_n P_2} \\
(\tau\text{-out}) & \dfrac{P \xrightarrow{?n} \nu\tilde{p}\langle P_1\rangle_n P_2}{n[P] \xrightarrow{\tau} \nu\tilde{p}(n[P_1]||P_2)} & \text{(Open)} & \dfrac{P \xrightarrow{\overline{open}} P'}{n[P] \xrightarrow{\overline{open}\langle n\rangle} P'} \\
(\tau\text{-open}) & \dfrac{P \xrightarrow{open\langle n\rangle} P' \quad Q \xrightarrow{\overline{open}\langle n\rangle} Q'}{P|Q \xrightarrow{\tau} P'|Q'} & \text{(Par)} & \dfrac{P \xrightarrow{\alpha} O \quad \alpha \neq exit\langle n,m\rangle}{P|Q \xrightarrow{\alpha} O|Q} \\
\text{(Par exit)} & \dfrac{P \xrightarrow{exit\langle n,m\rangle} \nu\tilde{p}\langle P_1\rangle_n P_2}{P|Q \xrightarrow{exit\langle n,m\rangle} \nu\tilde{p}\langle P_1|Q\rangle_n P_2} & \text{(Res)} & \dfrac{P \xrightarrow{\alpha} O \quad n \notin n(\alpha)}{(\nu n)P \xrightarrow{\alpha} (\nu n)O}
\end{array}
$$

$$
(\tau\text{-in}) \; \frac{P \xrightarrow{enter\langle n,m\rangle} \nu\tilde{p}\langle P_1\rangle_n P_2 \quad Q \xrightarrow{\overline{enter}\langle n,m\rangle} \nu\tilde{q}\langle Q_1\rangle_n Q_2}{P|Q \xrightarrow{\tau} \nu\tilde{p}\nu\tilde{q}(n[P_1|Q_1]||P_2|Q_2)} \qquad (\tau\text{-Amb}) \; \frac{P \xrightarrow{\tau} Q}{n[P] \xrightarrow{\tau} n[Q]}
$$

$$((fn(P_1) \cup fn(P_2)) \cap \{\tilde{q}\} = (fn(Q_1) \cup fn(Q_2)) \cap \{\tilde{p}\} = \emptyset)$$

Table 1: Labelled Transition System.

cidence between the reduction semantics and the labelled transition semantics, i.e. they describe the same process behavior from two different angles.

Theorem 1 *If* $P \to Q$ *iff* $P \xrightarrow{\tau}\equiv Q$.

3 Bisimulation Relation

In this section, we will provide a bisimulation relation based on the labelled transition system provided in Section 2. Note that in the transition system, for any transition $P \xrightarrow{\alpha} O$, O is either a process or a concretion. If we use this system directly, then we will have to give the definition for equivalence of two concretions, which is rather difficult and actually unnecessary. Our solution is inspired by [6, 7], which lies in that for each high order action whose result is concretion, such as $enter\langle n, m\rangle$, $\overline{enter}\langle n, m\rangle$, $exit\langle n, m\rangle$ and $?n$, we replace it with a corresponding *first order* action, which makes the result of transition to be a simple process. Thus a neat definition for bisimulation relation can be obtained by such a conversion from the high-order action to the first-order action.

Definition 2 $(\nu\tilde{p}\langle P\rangle_n Q) * R \overset{def}{=} \nu\tilde{p}(n[P|R]||Q)$, *where* $\{\tilde{m}\} \cap fn(R) = \emptyset$.

Definition 3 *(i) If* $P \xrightarrow{\alpha} K$, *then* $P \xrightarrow{\alpha R} K * R$, *where* $\alpha \in \{enter\langle n, m\rangle, exit\langle n, m\rangle\}$.

(ii) If $P \xrightarrow{\overline{enter}\langle n,m\rangle} R$, *then* $P \xrightarrow{\overline{enter}\langle n,m\rangle m[K]} K * m[R]$.

(iii) If $P \xrightarrow{?n} \nu\tilde{p}\langle P_1\rangle_n P_2$, *then* $P \xrightarrow{?} \nu\tilde{p}(n[P_1]||P_2)$

In fact, the above definition defines the transition between processes, then we can give the bisimulation relation in some standard form. Note the the definition in this paper is in the weak form.

Definition 4 ***(Ambient Bisimulation)*** *A binary symmetric relation S is called an ambient bisimulation if PSQ and $P \overset{\alpha}{\rightarrow} P'$, then Q' exists such that $Q \overset{\hat{\alpha}}{\Rightarrow} Q'$ and $P'SQ'$. Let $P \approx Q$ if there exist some ambient bisimulation S, such that PSQ.*

In general, for an equivalence relation on processes, we hope it is a congruence relation, that is, it should be closed under any operator. In fact, ambient bisimulation is a congruence relation, which shows the soundness of the definition in some sense.

Theorem 5 *Ambient bisimulation is a congruence relation.*

4 Proof for Algebra Laws

In this section, we will give neat proofs for some process equivalence laws, which are mainly taken from [4]. Different from them, the laws discussed in this section are not typed and due to the difference of the calculus, the form of process expression is modified accordingly. It is not easy to see that these algebra laws hold under new equivalence criterions and comparing to the complex contextual reasoning, the proof presented in this section is much neater, which shows that in some sense, for ROAM, as process equivalence criterion, the ambient bisimulation is more suitable.

Theorem 6 *The following laws hold:*

(i) $(\nu n)(n[\overline{in}\langle m\rangle.P] \| m[in\langle n\rangle.Q]) \approx (\nu n)n[P|m[Q]]$.

(ii) $(\nu m)(n[\overline{in}\langle m\rangle.P] \| m[in\langle n\rangle.Q]) \approx (\nu m)n[P|m[Q]]$.

(iii) $h[n[\overline{in}\langle m\rangle.P] \| m[in\langle n\rangle.Q] | \overline{out}\langle k\rangle.R]) \approx h[n[P|m[Q] | \overline{out}\langle k\rangle.R]$.

(iv) $h[n[\overline{in}\langle m\rangle.P] \| m[in\langle n\rangle.Q] | open\langle k\rangle.R]) \approx h[n[P|m[Q] | open\langle k\rangle.R]$.

(v) $n[\overline{out}\langle m\rangle.P|m[out\langle n\rangle.Q]] \approx n[P] \| m[Q]$.

(vi) $(\nu n)(open\langle n\rangle.P|n[\overline{open}.Q]) \approx (\nu n)(P|Q)$.

(vii) $h[open\langle n\rangle.P] \| n[\overline{open}.Q]] \approx h[P|Q]$

(viii) $h[open\langle n\rangle.P|n[\overline{open}.Q] | \overline{out}\langle k\rangle.R] \approx h[P|Q|\overline{out}\langle k\rangle.R]$.

(ix) $h[open\langle n\rangle.P|n[\overline{open}.Q] | open\langle k\rangle.R] \approx h[P|Q|open\langle k\rangle.R]$, *if* $k \neq n$.

Proof: We give a uniform proof for clause (i)-(ix). Let LHS and RHS denote the left and right hands of the formula respectively. Let

$$S = \{(LHS, RHS)\} \cup \approx$$

then it is easy to show S is an ambient bisimulation. We have done. □

References

[1] L.Cardelli, A.Gordon. Mobile Ambients. Theoretical Computer Science, 240(1), pp.177-213, 2000.

[2] X.Guan, Y.Yang, J.You. Typing Evolving Ambients. Information Processing Letters, 80(5), pp.265-270, Nov.2001.

[3] R.Milner, J.Parrow, D.Walker. A Calculus of Mobile Process, part I/II. Journal of Information and Computation, 100:1-40 and 41-77, Sept.1992.

[4] F.Levi, D.Sangiorgi. Mobile safe ambients. ACM Trans. Program. Lang. Syst. 25(1):1-69, 2003.

[5] Luca Cardelli, D. Gordon, Types for mobile ambients. Proc. of the 26th ACM SIGPLANSIGACT symposium on Principles of programming languages, San Antonio, Texas, USA, pp. 79 - 92, 1999.

[6] D.Sangiorgi. Expressing Mobility in Process Algebras: First-Order and Higher-Order Paradigms. PhD thesis, Department of Computer Science, University of Edinburgh, 1992.

[7] D.Sangiorgi. Bisimulation for Higher-Order Process Calculi. Information and Computation, 131(2):141-178, 1996.

Brill Academic Publishers
P.O. Box 9000, 2300 PA Leiden,
The Netherlands

Lecture Series on Computer and Computational Sciences
Volume 2, 2005, pp. 37-41

Modelling Heat Transfer in Work Rolls of a Continuous Hot Strip Mill. Part II: Operating Conditions

R. L. Corral[1*], R. Colás[2], A. Pérez[3]

[1] Instituto Tecnológico de Cd. Cuauhtémoc, Cuauhtémoc, Chihuahua, México:
[2] Facultad de Ingeniería Mecánica y Eléctrica, Universidad Autónoma de Nuevo León, San Nicolás de los Garza, NL, México
[3] Centro de Investigación en Materiales Avanzados, Chihuahua, México

Received 9 March, 2005; accepted in revised form 15 April, 2005

Abstract: The heat transfer phenomenon in work rolls, used for hot rolling steel strip, has been modeled following two different approaches; the first integrating the heat flow, while the second was done integrating the temperature by means of the complex combination method, which lead at Bessel's Modified Equation, and this, is rewritten in terms of Thomson functions. Both models were developed to compute the temperature distribution at steady-state in work rolls in a hot rolling mill, which are subject to successive heating and cooling cycles, obtaining as a result, a positive heat influx. The caused fatigue, it can originate some damage modes in the rolls, for that reason it is pertinent to know the thermal response in operating conditions to give them an appropriate maintenance to achieve effectiveness during the process. Besides, the models here presented, can be also employed to evaluate the effects that changes in operating conditions might cause, such as, the rolling speed, disposition of the cooling headers and thermal diffusivity effects. These approaches provides for a novel insight when certain rolling conditions are required in the industry.
Keywords: Mathematical modeling; Hot rolling; Heat transfer; Work rolls; Thermal diffusivity; Cooling headers; Velocity speed

Mathematics SubjectClassification: 80A20 Heat and mass transfer, heat flow

PACS: 44. Heat Transfer

1. Introduction

Different models have been developed in order to compute the thermal response in work rolls [1-3]. Here it is developed a model integrating the heat flow at the surface of the roll, and another for compute the thermal profile by means of special functions, both models during steady-state condition. One of the advantages of these models is that they allow to observe the response from the roll to the modification of some operation parameter before carrying out some change in the production line, that which is of great utility in the taking of operative decisions and of design. For example, if it is considered the option of increasing the production of steel strip modifying the speed of turn of the roll, or the relocation of the cooling headers or the use of a new roll type [3-7].

2. Description of the models

The models assume that the roll is heated when the strip is being deformed, and it is being cooled between one and other strip [1]. The temperature is computed in any point by the effect caused by one turn of the work roll, by the rolling of a strip and by the whole campaign. Both models assume that the thermal response can be expanded into a Fourier series. So, model I is expressed by:

* Corresponding author. Active member of Research National System. E-mail: evelynmarabel@yahoo.com.mx

$$f_\theta = \beta T_s^* \sum_{n=1}^{m} [A_n \cos(n\theta) + B_n \sin(n\theta)] \tag{1}$$

While, in model II, the sustained temperature is given by:

$$T = T_a + T_s^* A_0 + T_s^* \sum_{n=1}^{m} \left[\begin{array}{l} \frac{\ker(0,r_i) + \ker(0,r) + kei(0,r_i) kei(0,r)}{\ker^2(0,r_i) + kei^2(0,r_i)} A_n \cos\left(n\theta - \alpha\sqrt{n}\right) + \\ \frac{kei(0,r_i) + \ker(0,r) - kei(0,r)\ker(0,r_i)}{\ker^2(0,r_i) + kei^2(0,r_i)} B_n \sin\left(n\theta - \alpha\sqrt{n}\right) \end{array} \right] \tag{2}$$

where, f_θ is the angular variation of the heat flow, T_a is the ambient temperature, T_s* is the temperature at the surface, A_0, A_n and B_n are constants to be found, and α and β are a parameters which depends on the physical properties and angular velocity of the roll:

$$\alpha = \sqrt{\omega / 2\kappa r} \tag{3}$$

$$\beta = \sqrt{\rho c_p k \omega} \tag{4}$$

where ω is the angular velocity.

1. Results and discussion

The thermal profile for work rolls and the effects in changes in operating conditions were predicted by both models in similar form. The thermal response during one cycle becomes important in the surface of the roll, as it has been reported elsewhere [1-4], see Fig. 1.
Figure 2 shows the evolution of the average temperature in the work roll as effect of the multiple thermal turns, for the four different speeds and to oneself radial position. It is observed increase in the temperature with an increase of speed. Figure 3 shows the effect of the disposition of the cooling headers with respect to the work rolls. The cooling headers are of great importance, since with them the propagation of the thermal fracture in the work roll it is minimized [5] and the thermal expansion it is controlled [4,6,7]. Both models predicts that the temperature of the roll it is increased, when area affected by the cooling system, moves away of the deformation area. Similar results are in Patula [8], which nor consider the contact among the work roll and back-up roll, neither the free convection phenomenon as in this work. Starting from these studies it is known that to greater efficiency of the cooling system, the heating of the roll will be smaller.

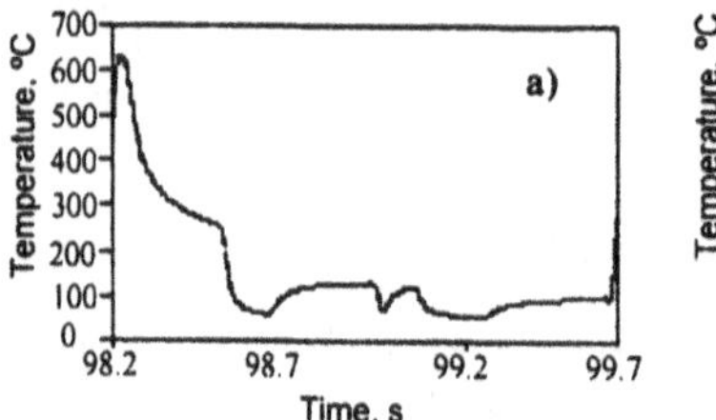

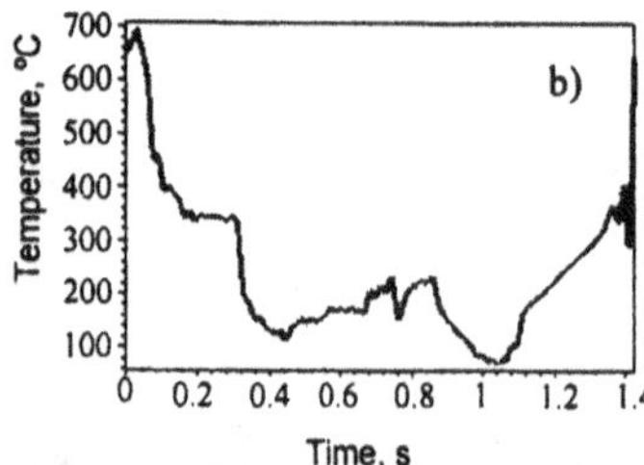

Figure 1: Thermal profile for a work roll during one cycle: a) model I, b) model II.

The mathematical model allows the simulation of the profile of temperatures of the roll as variation of its physical properties (see Table1). Is the attention centered in the thermal diffusivity (κ), maintaining constant the remaining parameters with the purpose of comparing the effects of this variation.

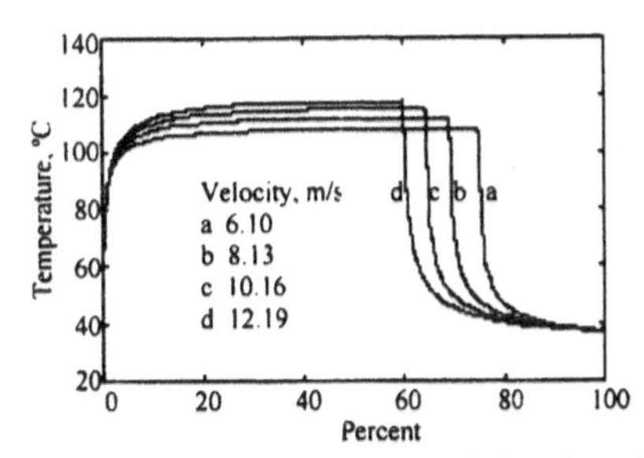

Figure 2: Increase of the temperature during the rolling of a strip and the period of wait of the following one.

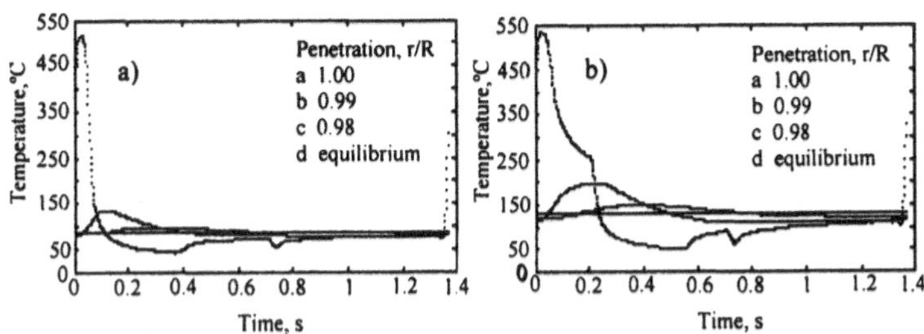

Figure 3: Temperature profile moving away the cooling system of the deformation area: a) 0°, and b) 45°

Table 1: Physical properties of work rolls and back-up roll.

Properties	Work roll A	B	C	D	Back up roll	Units
Conductivity, k	17.96	24.20	33.70	48.00	33.70	W m^{-1}°C^{-1}
Density, ρ	7581.00	7028.00	7137.00	7400.00	7137.00	Kg m^{-3}
Specific heat, c_p	463.00	520.00	476.00	520.00	476.00	J $Kg^{-1}K^{-1}$
Diffusivity, κ	5.19	6.62	9.90	12.47	9.90	$10^{-6}m^2s^{-1}$
Fig. 4	a and a'	b and b'	c and c'	d and d'		

In the Fig. 4 it is shown the effect of the thermal difussivity during one turn of the work roll in a, b, c, and d; and during the time the strip is deformed and the idle time between strips in a a', b', c', and d' for four different thermal diffusivities. The roll of greatest thermal difussivity presents a better definition of the changes that this experiment in its surface (Fig. 4 d and d').

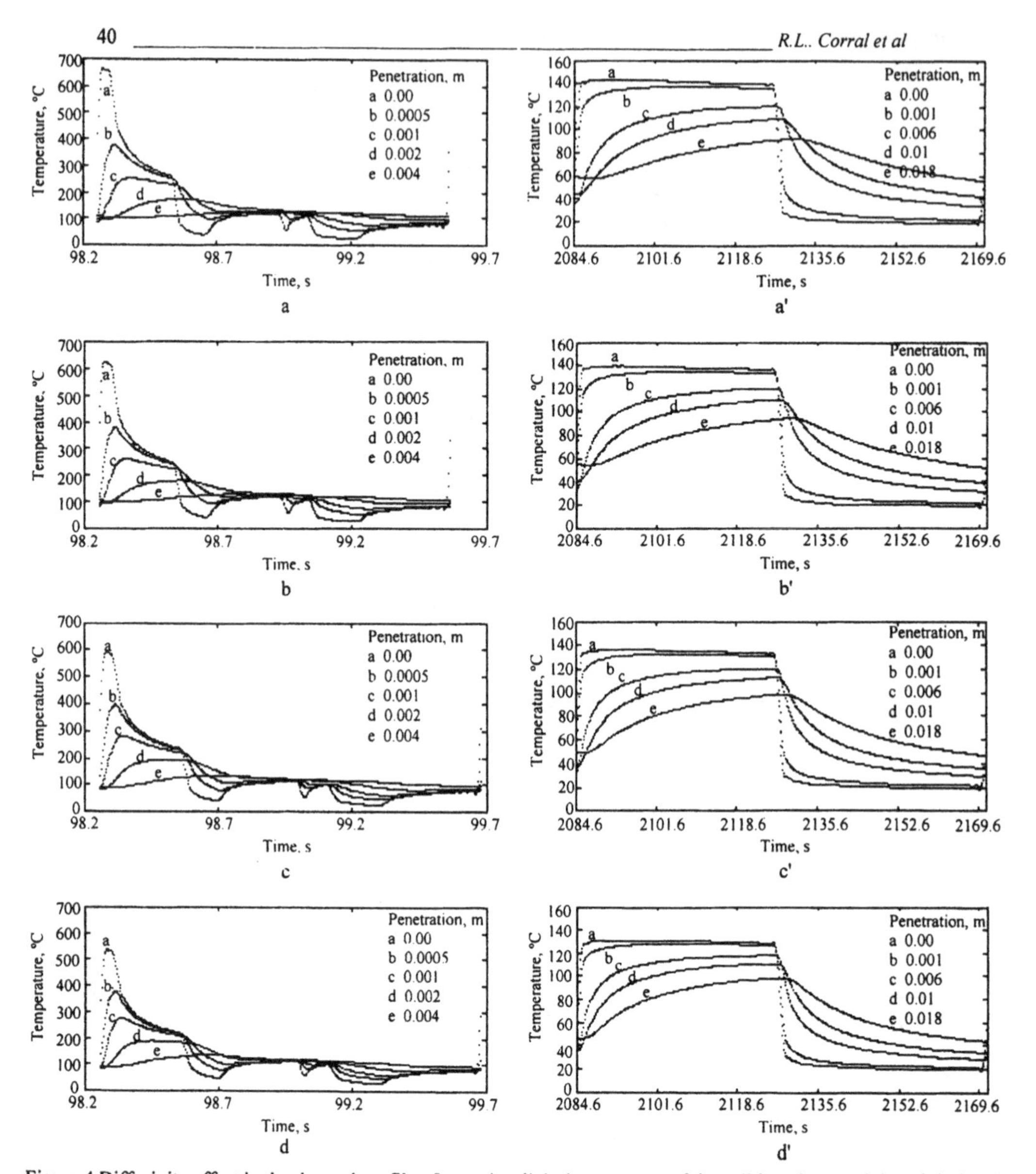

Figure 4 Diffusivity effect in the thermal profile of a work roll during one turn of the roll in a, b, c, and d; and during the time the strip is deformed and the idle time between strips in a a', b', c', and d': a and a' $5.19\times10^{-6}m^2s^{-1}$, b and b' $6.62\times10^{-6}m^2s^{-1}$, c and c' $9.90\times10^{-6}m^2s^{-1}$, and d and d' $12.47\times10^{-6}m^2s^{-1}$.

Referentes

[1] A. Pérez: Simulación de la temperatura en estado estable-dinámico para un rodillo de trabajo de un molino continuo de laminación en caliente. Doctoral Dissertation, Universidad Autónoma de Nuevo León, San Nicolás de los Garza, NL, México, 1994 (in Spanish).

[2] M.P. Guerrero, C.R. Flores, A. Pérez and R. Colás, Modelling heat transfer in hot rolling work rolls, *Journal of Materials Processing Technology*, 94, 52-59(1999).

[3] A. Pérez, R.L. Corral, R. Fuentes and R. Colás, Computer simulation of the thermal behaviour of a work roll during hot rolling of steel strip, *Journal of Materials Processing Technology* 153-154, 894-899(2004).

[4] R.L. Corral, R. Colás and A. Pérez, Modelling the thermal and thermoelastic responses of work rolls used for hot rolling steel strip, *Journal of Materials Processing Technology* 153-154, 886-893(2004).
[5] P.G. Stevens, K.P. Ivens and P. Harper, *Journal Iron and Steel Institute* 209 1-11(1971).
[6] W.L.Roberts, *Hot Rolling of Steel.* M. Dekker, Inc. New York, 1983.
[7] S.W. Wilmotte and J. Mignon, *C.R.M.* 34, 17(1973).
[8] Patula, E. J., *ASME Journal of Heat Transfer*, 103 36-41(1981).

Brill Academic Publishers
P.O. Box 9000, 2300 PA Leiden,
The Netherlands

Lecture Series on Computer and Computational Sciences
Volume 2, 2005, pp. 42-45

Singular Value Decomposition (SVD) Based Attack on Different Watermarking Schemes

Milind Engedy, Munaga.V.N.K.Prasad[1] and Ashutosh Saxena

IDRBT, Castle Hills, Road No 1, Masab Tank, Hyderabad, India.
milinde@mtech.idrbt.ac.in, {mvnkprasad, asaxena}@idrbt.ac.in

1. Introduction

The advancement in the field of digital technology and communication networks has led to unauthorized access and piracy of the digital documents over the Internet. A general belief is that preventing illegal copying of digital documents is difficult if not impossible. Digital Watermarking is an excellent tool at the disposal of the owners of the digital content to protect their Intellectual Property Rights (IPR). The digital documents can be audio, video, images, softwares etc. In this paper the digital documents we concentrated on, are images.

Digital watermarking is defined as a process by means of which a watermark dependent *pattern* is used to modify the host image intended for protection. The modifications made do not degrade the perceptual quality of the host image. The watermark can be a random number sequence or logo image of a company. The watermark is either visible or invisible. The watermarking techniques can be classified into: spatial and transform domain schemes. The watermarked image is called as cover image. The primary requirement of any digital watermarking schemes is robustness to the signal processing and intentional attacks. However the importance of each attack depends on the applications for which the watermarking is being used. There are a numerous watermarking schemes proposed in the literature [2,3], unfortunately very few of them focused on the requirements of robustness to intentional attacks. So far the intentional attacks on the watermarking schemes mainly concerned with either counterfeit of the watermark or removal/masking of the watermark.

In [1] Craver discussed a hypothetical situation where multiple claims of ownership of the cover image are possible i.e. the true owner of the watermarked image cannot claim the ownership based on the hidden watermark because the attacker can also provide sufficient evidence of his ownership over the cover image. Non-invertibility of the watermark function is presented as the essential condition to invalidate the multiple claims of ownership. In this paper we prove that non-invertibility property of the watermark function is not the alone condition to resolve rightful ownership of the image, but propose that the two ways one-to-one mapping between the watermark and *pattern* is also a necessary condition.

It is not necessary for the attacker to counterfeit the hidden watermark to claim his ownership since the legal authority doesn't have any knowledge about the watermark. Moreover the attacker need not go by the same algorithm as used by the owner as long as he can provide foolproof evidence of his ownership. Assume Alice, the owner of the digital images and Bob as one of the customer. To protect her Intellectual Property Rights (IPR) Alice would mark her images with the watermark dependent *pattern* and sell to Bob, a malicious user. Now suppose that the *pattern* would give rise to different images, among which the watermark used by Alice is one. Bob would extract the *pattern* and construct his own watermark due to one-to-many relationship and claim his ownership leading to the controversy over rightful ownership. Even the non-invertibility condition of the watermarking function could not protect the IPR of Alice.

SVD based watermarking scheme is the base for this attack since we take the advantage of the one-to-many relationship between the singular value matrix and the image matrix. In this paper we demonstrate this attack on 3 different watermarking schemes i.e. LSB substitution, CDMA spread spectrum watermarking and SVD based watermarking scheme.

[1] Corresponding Author

2. Attack Using SVD Based Watermarking and the Analysis

In SVD based watermarking every image matrix I is decomposed into the product of three matrices $I=USV^T$ where U and V are orthogonal matrices and S is a diagonal matrix with which contains the singular values of I, the columns of U are called left singular vectors of I and columns of V are called right singular vectors of I. The singular matrix S of the host image is modified to embed the watermark [5]. Each singular value specifies the luminance of the image whereas the corresponding pair of matrices specifies the geometry of the image [4].
The singular matrix S is unique to the given image I but the inverse may not be true since SVD is the one-to-many mapping, there exists many matrices whose S matrix may be the same [5]. The matrix S may be used in constructing another image I' such that

$$I' = U'SV'^T$$

The attack functions as follows:

$$X = E(I, w)$$

X is the watermarked image, E is watermarking function applied to image I and w is the watermark *pattern.*
The image X is available with Bob and he would decompose X to Singular values matrix

$$X = USV^T$$

The watermark possessed by Bob is y and he applies the singular value decomposition to the watermark

$$y = usv^T$$

The size of the watermark y can be different from X but for simplicity we chose size of y same as that of X. Bob would claim the ownership by simply replacing the S matrix in the place of s to reconstruct his watermark. Bob can also use any arbitrary relation of the type $\lambda_{wi}^{d*} = \lambda_i^{d*} - \lambda_i/\alpha$ to compute the singular matrix only to show that the process is inverse of the embedding operation.

$$y' = uSv^T$$

he would prove a very high correlation between the watermark y and reconstructed watermark y'. It would be difficult for the legal authorities to invalidate his claims, as Bob did not perform any modification to the image that would degrade the perceptual quality of the watermarked image.

$$\left.\begin{array}{l} Corr\,(y', w) \approx 0 \\ Corr\,(y, y') \approx 1 \end{array}\right\} \qquad (1)$$

The relation (1) support the Bob's claim of using his own authentic watermarking algorithm. Bob can also prove the existence of his watermark in the original image I possessed by Alice. The attack performed on different watermarking schemes takes the advantage of the fact that there are no standardized watermarking algorithms with legal authorities and registered watermarks.

3. Experimental Results

The attack is performed on CDMA (Code Division Multiple Access) spread spectrum watermarking scheme; LSB (Least Significant Bit) substitution algorithm and SVD based watermarking algorithm.
The watermarked host image (LSB substitution, CDMA, SVD) is transformed to wavelet domain and singular value decomposition is applied to the first level LL band. Similarly the watermark is wavelet transformed and the SV matrix is computed for the LL band wavelet coefficients of the watermark. The

SV matrix of the watermark wavelet coefficients is replaced with the cover image wavelet coefficients singular value matrix. The watermark coefficients are reverse transformed to obtain the reconstructed watermark. The correlation values obtained are 0.971, 0.975, 0.977 for LSB substitution, CDMA spread spectrum and SVD watermarking respectively. The attack was also successfully implemented in spatial domain with correlation values 0.939, 0.971, 0.941. The cover images are shown in the figure 1 and the reconstructed attackers watermarks are shown in figure 2.

4. Conclusion

We proved that one-to-one mapping between watermark and the *pattern* is a must condition to resolve the disputes involved due to multiple claims of rightful ownership. As there are no guidelines available with the legal authorities regarding the standard practices of watermarking, these sorts of attacks are obvious to happen. Moreover the registration of the watermark is also a pre-requisite condition because an attacker may show whatever is feasible as watermark; hence guidelines for images to be qualified as watermark must be specified. We recommend the standardization of the watermark schemes and the registration of the watermark as major step towards making the process of owner authenticity transparent.

(a) (b) (c)

Figure 1: (a) LSB watermarked image, (b) CDMA watermarked image (c) SVD watermarked image

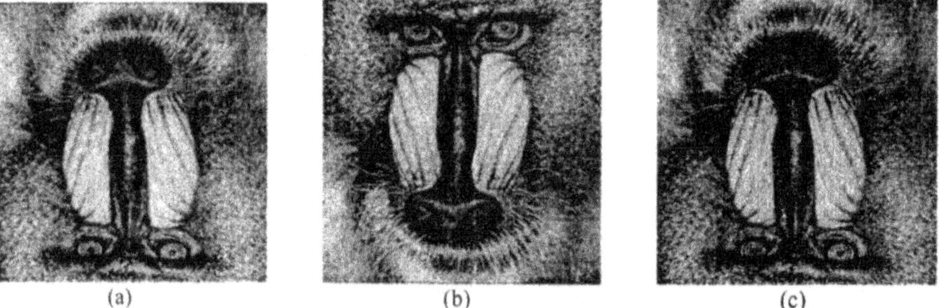

(a) (b) (c)

Figure 2: Reconstructed watermark from (a) LSB substitution, (b) CDMA watermarking, (c) SVD watermarking

References

[1] S. Craver, N. Menon, B. Yeo, and M. M Yeung, "Can Invisible watermarks resolve rightful ownerships?," in IS&T/SPIE Electronic Imaging: Storage and Retrieval of Image and Video Databases, vol 83 (6), pages 310-321,Feb.1997. L.D. Landau and F.M. Lifshitz: *Quantum Mechanics*. Pergamon, New York, 1965.

[2] I. J. Cox, J. Kilian, T. Leighton, T. Shamoon, "Secure spread spectrum watermarking for multimedia," IEEE Trans on Image processing vol 6 pp: 1673-1687, Dec 1997.G. Herzberg, *Spectra of Diatomic Molecules*, Van Nostrand, Toronto, 1950.

[3] D. Kundur, D.Hatzinakos, "Digital watermarking using multiresolution wavelet decomposition," Proceedings of the IEEE international conference on acoustics, speech and signal processing, Seattle, 1998.

[4] D. V. S. Chandra, "Digital Image Watermarking Using Singular Value Decomposition," Proceedings of 45th IEEE Midwest Symposium on Circuits and Systems, Tulsa, OK, August 2002, pp. 264-267.).

[5] R. Liu and T. Tan, "A SVD-Based Watermarking Scheme for Protecting Rightful Ownership," IEEE Transactions on Multimedia, 4(1), March 2002, pp.121-128.

Brill Academic Publishers
P.O. Box 9000, 2300 PA Leiden,
The Netherlands

Lecture Series on Computer and Computational Sciences
Volume 2, 2005, pp. 46-50

Updating TULV Decomposition

H. Erbay [1]

Mathematics Department,
Arts and Sciences Faculty,
Kırıkkale University,
Yahşihan, Kırıkkale 71100 Turkey

Received 5 April, 2005; accepted in revised form 22 April, 2005

Abstract: A truncated ULV decomposition (TULVD) of an $m \times n$ matrix A of rank k is a decomposition of the form $A = U_1 L V_1^T + E$, where U_1 and V_1 are left orthogonal matrices, L is a lower triangular matrix and E is an error matrix. We present an updating algorithm of order $O(nk)$ that reveals the rank correctly and produces good approximation to the subspaces of the matrix A.

Keywords: ULV decomposition, rank estimation, subspace estimation.

Mathematics Subject Classification: 65F25, 65F15

1 Introduction

Let A be an $m \times n$ matrix. A truncated ULV decomposition (TULVD) of A is of the form

$$A = U_1 L V_1^T + E \tag{1}$$

where for some $k \leq n$, $U_1 \in \Re^{m\times k}$, $V_1 \in \Re^{n\times k}$ are left orthogonal, that is, $U_1^T U_1 = V_1^T V_1 = I_k$, $L \in \Re^{k\times k}$ is nonsingular and lower triangular. and $E \in \Re^{m\times n}$ is an error matrix. The matrices L and E satisfy

$$\|L^{-1}\|_2 \leq \epsilon^{-1}, \quad \|E\|_2 \leq \epsilon, \quad U_1^T E = 0 \tag{2}$$

where ϵ is some tolerence.

Updating is a process that given the TULVD of A produce the TULVD of

$$A_u = \begin{pmatrix} \mathbf{a}^T \\ A \end{pmatrix}. \tag{3}$$

In this paper we present an updating algorithm that assumes U_1, L, and V_1 are stored, E is not stored. However, updating algorithms given by Stewart [12], Yoon and Barlow [1, 13] and Barlow, Erbay and Zhang [5] store and use the error matrix E. Thus, unlike these algorithms, our algorithm is suitable for large scale applications of low rank [8]. Our updating algorithm is most efficient when $k \ll n$. Numerical results show that our algorithm reveals the rank and also produces approximations to the subspaces of the matrix A.

2 TULVD Updating

An important subproblem in the updating algorithm is that given $\mathbf{a} \in \Re^n$ and $V_1 \in \Re^{n\times k}$ left orthogonal, find $\mathbf{v} \in \Re^n$ and $\mathbf{d} \in \Re^{k+1}$ such that

$$\|\mathbf{v}\| = 1, \qquad V_1^T\mathbf{v} = \mathbf{0}, \qquad \begin{pmatrix} V_1 & \mathbf{v} \end{pmatrix}\mathbf{d} = \mathbf{a}.$$

Such an algorithm is given in [4]. The algorithm runs in $O(nk)$ time.

Below are the steps of our updating algorithm.

Algorithm 1 (ULVD Updating Algorithm)

1. *Find* $\mathbf{v}$ *and* $\mathbf{d}$ *such that*

$$\mathbf{a} = (V_1 \quad \mathbf{v})\,\mathbf{d}, \quad \mathbf{d} = \begin{matrix} k \\ 1 \end{matrix}\begin{pmatrix} \mathbf{f} \\ \beta \end{pmatrix}.$$

Then

$$A_u = \begin{pmatrix} 0 & 1 \\ U_1 & 0 \end{pmatrix}\begin{pmatrix} L & 0 \\ \mathbf{f}^T & \beta \end{pmatrix}\begin{pmatrix} V_1^T \\ \mathbf{v}^T \end{pmatrix} + \begin{matrix} 1 \\ m \end{matrix}\begin{pmatrix} 0 \\ E \end{pmatrix}.$$

2. *Find the smallest singular value* σ_{k+1} *of the matrix* $\begin{pmatrix} L & 0 \\ \mathbf{f}^T & \beta \end{pmatrix}$ *such that*

$$\begin{pmatrix} L & 0 \\ \mathbf{f}^T & \beta \end{pmatrix}\mathbf{v}_{k+1} = \sigma_{k+1}\mathbf{u}_{k+1}.$$

3. ***If*** *$(\sigma_{k+1} \geq \epsilon)$* ***then*** *set*

$$\bar{L} = \begin{pmatrix} L & 0 \\ \mathbf{f}^T & \beta \end{pmatrix}, \quad \bar{U}_1 = \begin{pmatrix} 0 & 1 \\ U_1 & 0 \end{pmatrix}, \quad \bar{V}_1 = \begin{pmatrix} V_1^T \\ \mathbf{v}^T \end{pmatrix}, \quad \bar{E} = (I - \bar{U}\bar{U}^T)\,X_u.$$

else if *$(\sigma_{k+1} < \epsilon)$* ***then*** *find orthogonal matrices P, Q with P such that*

$$P^T\mathbf{u}_{k+1} = \mathbf{e}_{k+1}$$

and with Q such that

$$P^T\begin{pmatrix} L & 0 \\ \mathbf{f}^T & \beta \end{pmatrix}Q = \begin{pmatrix} \bar{L} & 0 \\ \bar{\mathbf{f}}^T & \beta \end{pmatrix}$$

maintains a lower triangular matrix.
Take

$$\begin{pmatrix} \bar{U}_1 & \bar{\mathbf{u}}_0 \end{pmatrix} = \begin{pmatrix} 0 & 1 \\ U_1 & 0 \end{pmatrix}P, \quad \begin{pmatrix} \bar{V}_1 & \bar{\mathbf{v}}_{k+1} \end{pmatrix} = \begin{pmatrix} V_1 & \mathbf{v}_{k+1} \end{pmatrix}Q$$

4. *Finally,*

$$A_u = \bar{U}_1\bar{L}\bar{V}_1^T + \bar{E}, \qquad \bar{E} = (I - \bar{U}\bar{U}^T)\,A_u$$

The updating algorithm runs $O(nk)$ time but if the signal subspace is also required then the run time becomes $O(nk^2)$.

Note that if $\sigma_{k+1} \geq \epsilon$ we have that

$$\|\bar{E}\|_F^2 = \|E\|_F^2$$

but, if $\sigma_{k+1} < \epsilon$ then

$$\|\bar{E}\|_F^2 = \|E\|_F^2 + \sigma_{k+1}^2.$$

Thus, if the new data increases the rank of the then the conditions in (2) hold. However, no increase in rank destroys the rank revealing character of the matrix. In this case one or repeated use of deflation algorithm [4, 12] not only restores it but also refines subspaces and reduces the effect of the error matrix E.

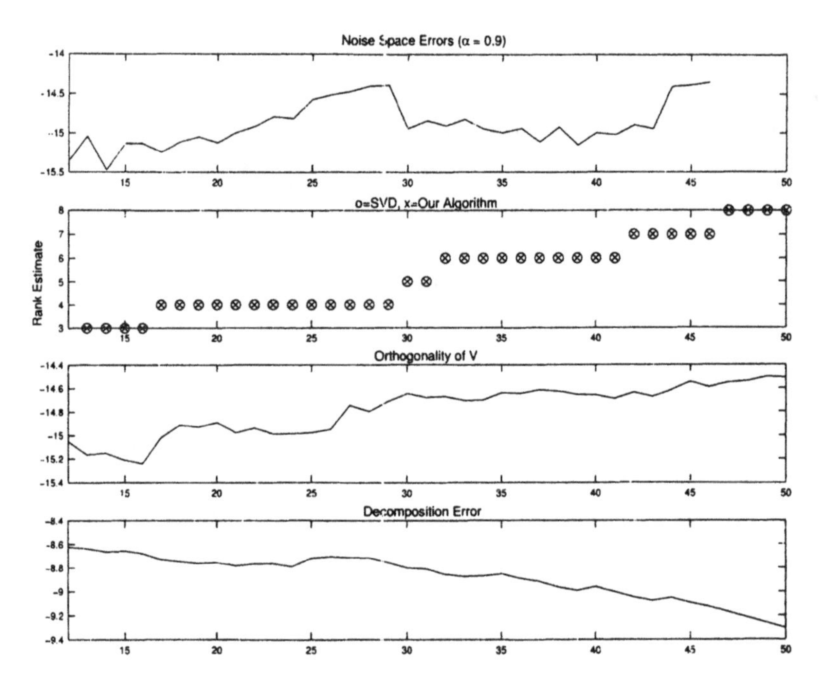

Figure 1: The plots show some result obtained by 1

3 Numerical experiments

The updating algorithm employed the so-called "exponential window" technique from signal processing. In this method, a value α, $0 < \alpha < 1$ which is called a forgetting factor, is multiplied to existing rows of the data matrix A before the new data $\mathbf{a}$ arrives, thus, the new data matrix becomes

$$A_u = \begin{pmatrix} \mathbf{a}^T \\ \alpha A \end{pmatrix}.$$

The tests were run in Matlab 6.5 on a personal compuer.

The MATLAB's SVD of the window matrix A was used as a reference in checking the accuracy. Let

$$A = X\,\Sigma\,Y^T, \qquad Y = (\; Y_1 \quad Y_2 \;)$$

be the SVD of A obtained by the Matlab's SVD algorithm. Also, let

$$A = U_1\; L\; V_1{}^T + E$$

be the TULVD of A obtained by our algorithm. In the Davis-Kahan [6] framework, the accuracy of the subspace errors are characterized by

$$\sin\theta = \|V_1^T Y_2\|. \tag{4}$$

We plotted $\sin\theta$ on the $\log 10$ scale to have better view of errors. We tracked the rank of the matrix A and compare it the Matlab's SVD. We checked the orthogonality of V_1 by computing

$$\|I_k - V_1^T V_1\|$$

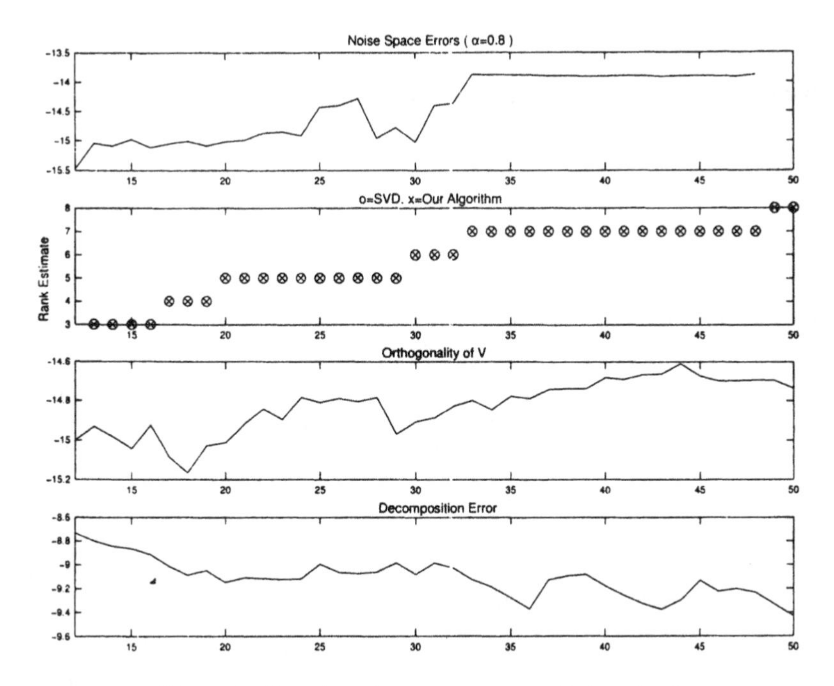

Figure 2: The plots show some result obtained by 2

and plotted. We also computed the decomposition error

$$\|A - U_1 L V_1\|.$$

Example 1 *B, 50-by-8 random matrix, chosen from a uniform distribution on the interval* $(0, 1)$. *42 randomly chosen rows of B multiplied by $\eta = 10^{-9}$ in order to vary the rank of the matrix, and $\epsilon \approx 10^{-8}$. The initial data matrix A of size 12-by-8 was chosen to be the first 12 rows of B. The new data* **a** *was chosen from the remaining rows of B at each window step in order. The forgetting factor* $\alpha = 0.9$.

Example 2 *B, 50-by-8 random matrix, chosen from a uniform distribution on the interval* $(0, 1)$. *42 randomly chosen rows of B multiplied by $\eta = 10^{-9}$ in order to vary the rank of the matrix, and $\epsilon \approx 10^{-8}$. The initial data matrix A of size 12-by-8 was chosen to be the first 12 rows of B. The new data* **a** *was chosen from the remaining rows of B at each window step in order. The forgetting factor* $\alpha = 0.8$.

The first plot in Figures (1) and (2) show the subspace errors obtained by our algorithm.

The second plot in Figures (1) and (2) show the rank estimates by our algorithm and MATLAB's SVD algorithm. Our algorithm estimates the rank correctly in spite of frequent rank changes. The horizontal axes represents the window steps.

4 Conclusion

We have presented an updating algorithm which is suitable for large scaled problems in which the rank of a matrix is small compared to the its size. Our numerical results show that the updating algorithm is promising.

References

[1] J.L. Barlow and P.A. Yoon, *Solving Recursive TLS Problems Using Rank-Revaling ULV Decomposition*, In S. Van Huf fel, editor, *Proc. Workshop on TLS and Errors-In-Variables*, Philadelphia, PA, SIAM Publications, 1997, pp. 117-126.

[2] J.L. Barlow P.A. Yoon and H. Zha, *An Algorithm And A Stability Theory For Downdating The ULV Decomposition*, BIT, 36(1996), 15-40.

[3] J.L. Barlow and H. Erbay, *Modifiable Low-Rank Approximation to a Matrix*, in preparation.

[4] J.L. Barlow, H. Erbay, and I. Slapničar, *Alternative Algorithms for Construction and Refinement of ULV Decompostions*, SIAM J. Matrix Anal. Appl., accepted for publication..

[5] J.L. Barlow H. Erbay and Z. Zhang, *A modified Gram-Schmidt based downdating technique for the ULV decompositions with applications to recursive TLS problems*, Adv. Sign. Proc. Alg., Arch. and Impl. IX, F.T. Luk, Editor, SPIE Proc., Bellingham, WA, 3807(1999), 247-257.

[6] C. Davis and W.M. Kahan, *The rotation of eigenvectors by a perturbation III*, SIAM J. Num. Anal., 7(1970), pp. 1-46.

[7] D.K. Faddeev V.N. Kublanovskaya and V.N. Faddeeva, *Solution Of Linear Algebraic Systems With Rectangular Matrices*, Proc. Steklov Inst. Math., 96(1968), pp. 93-111.

[8] R. Fierro L. Vanhamme and S. Van Huffel ,*Total Least Squares Algorithms Based On Rank-Revealing Complete Orthogonal Decompositions*, SIAM Publications, 1997, pp. 99-116.

[9] M. Gu and S. Eisenstat,*Downdating The Singular Value Decomposition*, SIAM J. Matrix Anal. Appl., 16(1995), pp. 793-810.

[10] R.J. Hanson and C.L. Lawson,*Extensions And Applications Of The Householder Algorithm For Solving Linear Least Squares Problems*, Math. Comp., 23(1969), pp. 787-812.

[11] B.N. Parlett, *The Symmetric Eigenvalue Problem*, SIAM Publications, 1997.

[12] G.W. Stewart,*Updating A Rank-Revealing ULV Decomposition*, SIAM J. Matrix Anal. Appl., 14(1993), pp. 494-499.

[13] P.A. Yoon and J.L. Barlow, *An Efficient Rank Detection Procedure For Modifying The ULV Decomposition* , BIT, 38(1998), pp. 781-801.

Brill Academic Publishers
P.O. Box 9000, 2300 PA Leiden,
The Netherlands

Lecture Series on Computer and Computational Sciences
Volume 2, 2005, pp. 51-54

Analysis of the Wavelet Transform Parameters in Images Processing

G. Farias[1], M. Santos[2]

[1] Dpto. de Informática y Automática
E.T.S.I. Informática – UNED
28040 Madrid, Spain

[2] Dpto. de Arquitectura de Computadores y Automática
Facultad de Informática, Universidad Complutense
28040 Madrid, Spain

Received 8 April, 2005; accepted in revised form 21 April, 2005

Abstract: Wavelet transform (WT) is a powerful computational tool that allows to map signals to spaces of lower dimensionality and to extract their main features. But the transform has many interconnected parameters that have a strong influence on the final results. In this paper, different configurations of wavelets are analyzed in order to optimize the classification and retrieval of two-dimensional signals. The study is applied to nuclear fusion plasma images. Results show how the classification varies when using one or another configuration.

Keywords: Wavelets transform, signal processing, classification.

1. Introduction

Databases in nuclear fusion experiments, as in other application areas, are made up of thousands of signals. Moreover, these signals are large and complex, and it is difficult to extract their main features without automatic mechanisms.

In [1] and [2] the *Discrete Fourier Transformation* (DFT) is proposed to find similar time sequences. But the DFT has some limitations when it is applied to fast varying waveforms: time information can be lost when transforming the signals and transitory characteristics can be undetected. *Wavelet Transform* (WT) offers an efficient alternative to data processing and provides some advantages as: data compression, computational efficiency, and simultaneous time and frequency representation.

For this reason wavelets are widely applied. But the wavelet transform has many parameters that have to be selected and its configuration will have a strong influence on the final results. In this paper several types of wavelet mothers with different decomposition levels and different approximation coefficients are tried to show their influence on the processing of two-dimensional signals.

In order to evaluate the different WT configurations, wavelets have been applied to plasma images that have been provided by the nuclear fusion environment TJ-II stellarator. These images are later classified by the Support Vector Machines (SVM) algorithm [3]. The influence of the WT parameters on the performance of the classification is shown.

2. Parameters of the Discrete Wavelet Transform

Many researchers recognised that multi-scale transformations are very useful in a number of image processing applications [4]. The use of the *Discrete Wavelets Transform* (DWT) makes possible to

[1] Corresponding author. E-mail: gfarias@bec.uned.es

reach a desired decomposition level preserving the image information. The redundant information is minimized and so the computational load is substantially cut down [5, 6].

The Wavelet transform works with the so-called Mother Wavelet which is a prototype function. The mother wavelet is scaled and shifted to be compared with the original signal. The DWT computes a few correlation coefficients for each signal in a fast way. From these coefficients it is possible to reconstruct the original signal by using the inverse of the Wavelet Transform [5, 7].

The 2-D wavelet analysis operation consists in filtering and down-sampling horizontally using the 1-D lowpass filter L and the highpass filter H to each row in the image $I(x,y)$, producing the coefficient matrixes $I_L(x,y)$ and $I_H(x,y)$. Subsequently filtering and down-sampling produces four subimages $I_{LL}(x,y)$, $I_{LH}(x,y)$, $I_{HL}(x,y)$ and $I_{HH}(x,y)$ for one level of decomposition. $I_{LL}(x,y)$ represents the coarse approximation of $I(x,y)$. $I_{LH}(x,y)$, $I_{HL}(x,y)$ and $I_{HH}(x,y)$ are detail subimages, which represent the horizontal, vertical and diagonal directions of the image $I(x,y)$.

Once the wavelet transform is applied for all-possible scales (or levels) and different mother wavelets, some characteristics of the signals are obtained. However, to select the most suitable family of mother wavelets and the best scale for particular signals is a difficult task.

3. Application Images

In order to evaluate the different wavelet approaches, several experiments have been carried out in the TJ-II Thomson Scattering device. Thomson Scattering (TS) diagnostic images measure the electronic density and radial temperature of the nuclear fusion stellarator. These two-dimensional can be grouped into the classes of Table 1.

Table 1: Description of TJ-II Thomson Scattering patterns

Pattern	Description
BKGND	CCD camera background
COFF	Reached cut off density for plasma heated by Electron Cyclotron Resonant Heating (ECRH)
ECH	Plasma heated by Electron Cyclotron Resonant Heating (ECRH)
NBI	Plasma heated by Neutral Beam Injectors (NBI)
STRAY	Parasitic light without plasma

As in any classification process, Thomson Scattering images need to be pre-processed in a suitable way [4]. Most of the analyses try to extract either unique or common signal features, allowing identification of patterns that reflect similar experimental conditions [9, 10]. The present work uses the Wavelet Transform to feature extraction and SVM for the classification.

4. Evaluation Parameters and Criteria

The evaluation must be done on an application as the important point when applying wavelets is to preserve the information for some purpose while reducing the signals size. During the experimentation, the parameters of the classifier and the hardware configuration remained fixed so as to evaluate only the influence of the wavelets parameters. The values that will be shown are mean values and they have been obtained after carrying out more than fifty experiments with each configuration.

The mother wavelet plays an important role in the final results. Its selection usually comes from experimentation. Other relevant factors to be considered are the features of the wavelet coefficients matrix and the decomposition level of the transform. Therefore in this paper the wavelets parameters that are varied to observe their influence are:

- **The family of Mother Wavelet**: The following are considered: Haar, DB2, DB3 and Dmey
- **Coefficients**: A: coarse approximation and V: vertical detail
- **Level or resolution**: 2, 4, 6 and others

On the other hand, the criteria that are going to be considered in order to evaluate the performance of the classification process for different wavelet parameters are:

- **Processing time**: Elapsed time to get the signal classified.
- **Hits rate or efficiency**: The percentage of success, i.e., the number of signals that have been classified correctly.
- **Classification power**: Ratio between the efficiency and the processing time.

5. Results

5.1 Haar Wavelet Transform

In relation to the TS two-dimensional signals, it has been found that the best coefficient to characterize the images is the vertical detail, when the selected Mother Wavelet is Haar at level 4. Table 2 shows how in this case (level 4, V) the efficiency reaches the maximum (89.17%) and the processing time is minimum. The classification power is 188.68.

Table 2: Wavelet Haar Transform results

Level		Approximation			Vertical Detail		
		P. Time	*Hits rate*	*Power*	*P. Time*	*Hits rate*	*Power*
	1	1.2466	84.4425	67.73824803	2.4725	61.946	25.05399393
	2	0.58663	84.997	144.8903056	0.80978	76.111	93.9897256
	3	0.48013	86.387	179.9241872	0.51088	86.665	169.6386627
	4	**0.47087**	86.1095	182.8731922	0.47259	**89.1655**	**188.674115**
	5	0.50083	84.443	168.6061139	0.49826	86.942	174.4912295
	6	0.53412	82.4995	154.4587359	0.53046	83.888	158.14199

5.2 DB 2 and DB 3 Wavelet Transforms

Table 3 shows the results when using order 2 Daubechies wavelet. For DB2 and DB3 cases the best result as for the classification power and the processing time is obtained when the decomposition level of the transform is 4, with vertical detail.

Table 3: DB2 Wavelet Transform results

Level		Approximation			Vertical Detail		
		P. Time	*Hits rate*	*Power*	*P. Time*	*Hits rate*	*Power*
	2	0,61987	85,275	137,569168	-	-	-
	3	0,52477	86,387	164,618785	0,55915	83,0545	148,537065
	4	0,51252	85,554	166,92812	**0,5099**	85,2765	**167,241616**
	5	0,53615	84,165	156,980323	0,53763	**88,887**	165,331176
	6	0,56862	81,944	144,110302	0,56724	82,7765	145,928531

5.3 Dmey Wavelet Transform

Table 4 summarizes the result of applying the Dmey transform to the classification process. The best results are now between level 2 and 3 for the coarse approximation in terms of processing time and classification power, and level 4 is the best regarding hits rate with the vertical detail coefficient.

6. Conclusions

As it has been shown, the parameters of the wavelet transform have a great influence on the final result of images processing applications. In this paper, several mother wavelets with different decomposition level have been tried to show how the classification of two-dimensional signals varies.

In terms of mother wavelets, even though Haar is the simplest, it is the one which gives better results. As for the coefficients, in most of the cases the vertical detail was better in order to maximize the efficiency and the classification power, whereas the course approximation performed better in terms of the processing time.

Table 4: Dmey Wavelet Transform results

Level		Approximation			Vertical Detail		
		P. Time	*Hits rate*	*Power*	*P. Time*	*Hits rate*	*Power*
	1	2,3661	84,72	35,8057563	-	-	-
	2	**1,9198**	84,7195	44,1293364	-	-	-
	3	1,9231	86,387	**44,920701**	2,0649	77,5005	37,532326
	4	1,9946	86,3875	43,3106889	2,0443	**87,7765**	42,9371912
	5	2,0938	84,165	40,197249	2,0974	87,2215	41,5855345
	6	2,2086	81,666	36,9763651	2,1755	84,1665	38,6883475

In any case, it is possible to state that the decomposition level reaches an optimum. Regarding the efficiency and the power of classification, in most of the signals the best level corresponds to level 4. Yet, the time processing presents the best results between level 3 and 4.

These results facilitate the application of this computational method. The search space of a suitable configuration for these kinds of images has been reduced. That is, if other mother wavelets are wanted to be tried, the analysis could be reduced to the vertical coefficient and decomposition levels around 4.

Acknowledgments

The authors wish to thank to the TJ-II team for the help and the images they have provided.

References

[1] D. Rafiei, A. Mendelzon, Efficient Retrieval of Similar Time Sequences Using DFT, *Proc. 5th Inter. Conf. on Foundations of Data Organization* (FODO'98), 249-257 (1998).

[2] H. Nakanishi, T. Hochin, M. Kojima, Search and retrieval methods of similar plasma waveforms. *4th IAEA TCM on Control, Data Acquisition, and Remote Participation for Fusion Research*. San Diego. July 21-23 (2003).

[3] V. Vapnik, *The Nature of Statistical Learning Theory*, 2nd Edition, Springer (2000).

[4] J.L. Starck, F. Murtagh, A. Bijaoui, *Image Processing and Data Analysis: The multiscale approach.* Cambridge, University Press (2000).

[5] I. Daubechies, *Ten Lectures on Wavelets*, SIAM, Philadelphia (1992).

[6] M. Vetterli, Wavelets, Approximation and Compression, *IEEE Signal Processing Magazine*, 18(5), 59-73 (2001).

[7] M. Santos, G. Pajares, M. Portela, J.M. de la Cruz, A new wavelets image fusion strategy. *Lecture Notes in Computer Science*. Springer-Verlag, Berlin, 2652, 919-926 (2003)

[8] C. Alejaldre, et. al., First plasmas in the TJ-II flexible Heliac, *Plasma Physics and Controlled Fusion*, 41, A539-A548 (1999).

[9] R. Duda, P. Hart, D. Stork, *Pattern Classification*, 2nd Edition, a Wiley-Interscience (2000).

[10] S. Dormido-Canto, G. Farias, R. Dormido, J. Vega, J. Sánchez, M. Santos, TJ-II wave forms analysis with wavelets and support vector machines, *Review of Scientific Instruments* 75(10), 4254-4257 (2004)

Brill Academic Publishers
P.O. Box 9000, 2300 PA Leiden,
The Netherlands

Lecture Series on Computer and Computational Sciences
Volume 2, 2005, pp. 55-58

The properties of the Zadeh extension of $\log_a x$

Wang Guixiang[1]

Science School, Hangzhou Dianzi University, Hangzhou, 310018, People's Republic of China
Information and Telecommunication Engineering School, Harbin Engineering University
Harbin, 150000, People's Republic of China

Li Youming

Science School, Agricultural University of Hebei, Baoding, 071001, People's Republic of China

Zhao Chunhui

Information and Telecommunication Engineering School, Harbin Engineering University
Harbin, 150000, People's Republic of China

Received 5 April, 2005; accepted in revised form 22 April, 2005

Abstract: In this paper, we obtain some properties of the Zadeh extension (is restricted to fuzzy number space E, and denoted by $\log_{\hat{a}} u$) of function $g(x) = \log_a x$ about continuity, monotonicity, convexity (or concavity), differentiability and integration.

Keywords: Fuzzy numbers; Fuzzy mappings; Zadeh extensions; Differentiability; Integration.

Mathematics Subject Classification: 04A72, 28E10

In 1972, S. S. L. Chang, L. A. Zadeh [1] introduced the concept of fuzzy numbers with the consideration of the properties of probability functions. Since then many results on fuzzy numbers have been achieved (see for example [2,3,4,5,8,9]) because of the wide usage in theory and applications. With the development of theories of fuzzy number and its application, the concept of fuzzy numbers becomes more and more important. The theory and applications of fuzzy numbers often involve the mappings (be called fuzzy number mappings in this paper) from the set of fuzzy numbers into itself (see for example [6,7,10,11]). In addition, it is well known that the concepts of continuity, monotonicity, convexity (or concavity) and differentiability are very important in mathematics, and $f(x) = \log_a x$ is very important elementary function. The purpose of this paper is to discuss the properties of the Zadeh extension (is restricted to fuzzy number space E, and denoted by $\log_{\hat{a}} u$) of function $g(x) = \log_a x$ about continuity, monotonicity, convexity (or concavity), differentiability and integration.

Let R be the real number line. Denote $E = \{u |\ u : R \to [0,1]$ is normal, fuzzy convex, upper semi-continuous and its support $[u]^0$ is a compact set$\}$, where $[u]^0 = cl\{x \in R|\ u(x) > 0\}$. For any $u \in E$, u is called a fuzzy number and E is called fuzzy number space.

Obviously, $[u]^r$ is nonempty bounded closed intervals (denoted $[\underline{u}(r), \overline{u}(r)]$) for any $u \in E$ and $r \in [0,1]$, where $[u]^r = \{x \in R|\ u(x) \geq r\}$ when $r \in (0,1]$.

For any $a \in R$, define a fuzzy number $\hat{a}$ by $\hat{a}(t) = \begin{cases} 1 & if \quad t = a \\ 0 & if \quad t \neq a \end{cases}$ for any $t \in R$.

[1]Corresponding author. Science School, Hangzhou Dianzi University, Hangzhou, 310018, People's Republic of China. E-mail: wangguixiang2001@263.net

The addition, scalar multiplication and multiplication on E^1 is defined by:

$$(u+v)(x) = \sup_{y+z=x} \min[u(y), v(z)], \quad (\lambda u)(x) = \begin{cases} u(\lambda^{-1}x) & if \quad \lambda \neq 0 \\ \hat{0} & if \quad \lambda = 0 \end{cases}$$

$$(uv)(x) = \sup_{yz=x} \min[u(y), v(z)]$$

for $u, v \in E, \lambda \in R$.

It is well known that for any $u, v \in E, \lambda \in R$, $u+v, \lambda u, uv \in E$ and $[u+v]^r = [u]^r + [v]^r, [\lambda u]^r = \lambda[u]^r$, and $[uv]^r = [u]^r[v]^r$ for every $r \in [0,1]$.

For $u, v \in E$, we define $u \leq v$ iff $[u]^r = [\underline{u}(r), \overline{u}(r)] \leq [v]^r = [\underline{v}(r), \overline{v}(r)]$ for any $r \in [0,1]$. And $[u]^r \leq [v]^r$ iff $\underline{u}(r) \leq \underline{v}(r)$ and $\overline{u}(r) \leq \overline{v}(r)$.

If for $u, v \in E$, there exists $w \in E$ such that $u = v + w$, then we say the (H)difference $u - v$ to exist, and denote $u - v = w$. It is obvious that if the (H)difference $u - v$ exists, then $\underline{(u-v)}(r) = \underline{u}(r) - \underline{v}(r)$, $\overline{(u-v)}(r) = \overline{u}(r) - \overline{v}(r)$.

For $u, v \in E$, define $D(u,v) = \sup_{r\in[0,1]} \max(|\underline{u}(r) - \underline{v}(r)|, |\overline{u}(r) - \overline{v}(r)|)$, then (E, D) is a complete metric space, and satisfies $D(u+w, v+w) = D(u,v)$ and $D(ku, kv) = |k|D(u,v)$ for any $u, v, w \in E$ and $k \in R$.

A mapping $F: E \to E$ is said to be a fuzzy number mapping. For any $r \in [0,1]$, denote $[F(u)]^r = [\underline{F(u)}(r), \overline{F(u)}(r)], u \in E$. A mapping $F: E \to R$ is said to be a fuzzy number functional.

Definition 1[6] *Let $F: E \to E$ (R) be a fuzzy number mapping (functional).*

(1) We call F a convex (resp. concave) fuzzy number mapping (functional) if for any $u, v \in E$, $t \in (0,1)$, $F(tu + (1-t)v) \leq tF(u) + (1-t)F(v)$ (resp. $F(tu + (1-t)v) \geq tF(u) + (1-t)F(v)$).

(2) We call F a increasing (resp. decreasing) fuzzy number mapping (functional) if for any $u, v \in E$ with $u \leq v$, $F(u) \leq F(v)$ (resp. $F(u) \geq F(v)$) holds.

Denote $F(u) = \log_{\tilde{a}} u$, $u \in E_{(0,+\infty)}$ is the Zadeh extension of function $f(x) = \log_a x$ $(a > 0, a \neq 1)$, $x \in (0, +\infty)$, where $E_{(0,+\infty)} = \{u \in E | \ [u]^r \subset (0,+\infty)\}$.

Proposition 1 *(1) As $a > 1$, $F(u) = \log_{\tilde{a}} u$ is an increasing and concave continuous fuzzy number mapping on $(E_{(0,+\infty)}, D)$ and satisfies $[\log_{\tilde{a}} u]^r = [\log_a \underline{u}(r), \log_a \overline{u}(r)]$ for any $r \in [0,1]$ and $u \in E_{(0,+\infty)}$.*

(2) As $0 < a < 1$, $F(u) = \log_{\tilde{a}} u$ is an decreasing and convex continuous fuzzy number mapping on $(E_{(0,+\infty)}, D)$ and satisfies $[\log_{\tilde{a}} u]^r = [\log_a \overline{u}(r), \log_a \underline{u}(r)]$ for any $r \in [0,1]$ and $u \in E_{(0,+\infty)}$.

Definition 2[7] *Let $F: (E, D) \to (E, D)$ be a fuzzy number mapping, $u_0 \in E$.*

(1) if for $v_0 \in E$, there exists $w_0^+ \in E$ such that

$$\lim_{h\to 0^+} \frac{D(F(u_0 + hv_0), F(u_0) + hw_0^+)}{h} = 0$$

then we say F to be right G-differentiable at u_0 with respect to direction v_0, and call the unique (the uniqueness can be proved) w_0^+ (denote $F'_+(u_0, v_0) = w_0^+$) the right G-derivative of F at u_0 with respect to direction v_0.

(2) If for $v_0 \in E$, there exists $\delta_0 > 0$ and $w_0^- \in E$ such that the (H)difference $u_0 - \delta_0 v_0$ exists, and

$$\lim_{h\to 0^+} \frac{D(F(u_0), F(u_0 - hv_0) + hw_0^-)}{h} = 0$$

then we say F to be left G-differentiable at u_0 with respect to direction v_0, and call the unique (the uniqueness can be proved) w_0^- (denote $F'_-(u_0, v_0) = w_0^-$) the left G-derivative of F at u_0 with respect to direction v_0.

(3) If F is right and left G-differentiable at u_0 with respect to direction v_0, and $F'_+(u_0, v_0) =$

$F'_-(u_0, v_0)$, *then we say* F *to be G-differentiable at* u_0 *with respect to direction* v_0, *denote* $F'(u_0, v_0) = F'_+(u_0, v_0) = F'_-(u_0, v_0)$, *and call* $F'(u_0, v_0)$ *the G-derivative of* F *at* u_0 *with respect to direction* v_0.

Proposition 2 *Let* $u_0 \in E_{(0,+\infty)}$ *and* $a \in (0,1) \cup (1, +\infty)$.

(1) If $v_0 \in E$ *satisfies that* $\frac{\underline{v_0}(r)}{\underline{u_0}(r)}$ *is nondecreasing,* $\frac{\overline{v_0}(r)}{\overline{u_0}(r)}$ *is nonincreasing, and* $\frac{\underline{v_0}(1)}{\underline{u_0}(1)} \leq \frac{\overline{v_0}(1)}{\overline{u_0}(1)}$, *then* $F(u) = \log_{\tilde{a}} u$ *is right G-differentiable at* u_0 *with respect to direction* v_0, *and*

$$[(\log_{\tilde{a}} u)'_{(u_0^+, v_0)}]^r = \frac{1}{\ln a}[\frac{\underline{v_0}(r)}{\underline{u_0}(r)}, \frac{\overline{v_0}(r)}{\overline{u_0}(r)}]$$

for any $r \in [0,1]$.

(2) If $v_0 \in E_{(0,+\infty)}$ *satisfies that* $\frac{\underline{v_0}(r)}{\underline{u_0}(r)}$ *is nondecreasing,* $\frac{\overline{v_0}(r)}{\overline{u_0}(r)}$ *is nonincreasing,* $\frac{\underline{v_0}(1)}{\underline{u_0}(1)} \leq \frac{\overline{v_0}(1)}{\overline{u_0}(1)}$, *and there exists* $\delta_0 > 0$ *such that the (H)difference* $u_0 - \delta_0 v_0$ *exists, then* $F(u) = \log_{\tilde{a}} u$ *is G-differentiable at* u_0 *with respect to direction* v_0, *and*

$$[(\log_{\tilde{a}} u)'_{(u_0, v_0)}]^r = \frac{1}{\ln a}[\frac{\underline{v_0}(r)}{\underline{u_0}(r)}, \frac{\overline{v_0}(r)}{\overline{u_0}(r)}]$$

for any $r \in [0,1]$.

Let $u, v \in E$, we say u is comparable with v if $u \leq v$ or $v \leq u$ holds.

Let $U \subset E, u_0 \in E$, if $u \leq u_0$ for all $u \in U$, or $u \geq u_0$ for all $u \in U$, then we say all $u \in U$ are on the same side of u_0.

Let $u_0, v_0 \in E$, $u_0 \neq v_0$. According to the order '$\prec$' (we call it line order) that $w_1 \prec w_2 \Longleftrightarrow t_1 \leq t_2$ (orders '$\prec$' and '$\leq$' are not same), the line segment $\{w \in E : w = (1-t)u_0 + tv_0, t \in [0,1]\}$ educed by u_0 and v_0 becomes a directed set, i.e. directed lines segment, where $w_i = (1-t_i)u_0 + t_i v_0$ $(i = 1, 2)$. We denote the directed line segment by $[u_0 v_0]$.

Definition 3 [10] *Let* $u_0, v_0 \in E$ *satisfy* $u_0 \neq v_0$ *and suppose that the (H)difference* $v_0 - u_0$ *exists, and let* $F : [u_0 v_0] \to E$ *be a fuzzy number mapping. Taking* n *arbitrary fuzzy numbers* $w_1, w_2, \cdots, w_n$ *such that* $w_i \in (u_0 v_0), w_i \neq w_{i-1}$ *and* $[w_{i-1} w_i] \cap [w_i w_{i+1}] = w_i, i = 1, 2, \cdots, n$, *where* $w_0 = u_0$ *and* $w_{n+1} = v_0$ *(denote* $\tilde{P} = \{w_0, w_1, \cdots, w_{n+1}\}$, *and call* $\tilde{P}$ *a partition of directed line segment* $[u_0 v_0]$*), we denote* $\tilde{\Delta}_i = w_i - w_{i-1}$ *(by taking method of* w_i, *we see that* $w_i \in (w_{i-1} v_0]$, *so we know that* $\tilde{\Delta}_i$ *exists). And taking arbitrarily* $\tilde{\xi}_i \in [w_{i-1} w_i]$, *we denote* $\tilde{\xi} = \{\tilde{\xi}_1, \tilde{\xi}_2, \cdots, \tilde{\xi}_{n+1}\}$, $\tilde{S}_{u_0,v_0}(\tilde{P}, \tilde{\xi}) = \sum_{i=1}^{n+1} F(\tilde{\xi}_i)\tilde{\Delta}_i$. *If as* $||\tilde{P}|| = \max_{1 \leq i \leq n+1} D(\hat{0}, \tilde{\Delta}_i) \to 0$, *the limit of* $\tilde{S}_{u_0,v_0}(\tilde{P}, \tilde{\xi})$ *exists (and is independent of the partition* $\tilde{P}$ *and the selected points* $\tilde{\xi}_i$*), i.e., there exists a fuzzy number which is independent of the partition* $\tilde{P}$ *and the selected points* $\tilde{\xi}_i$, *denoted by* $\int_{u_0}^{v_0} F(u)du$, *such that* $\lim_{||\tilde{P}|| \to 0} \tilde{S}_{u_0,v_0}(\tilde{P}, \tilde{\xi}) = \int_{u_0}^{v_0} F(u)du$, *then we say* F *to be integrable over directed line segment* $[u_0, v_0]$, *and the limit* $\int_{u_0}^{v_0} F(u)du$ *is called the integration of* F *over directed line segment* $[u_0, v_0]$.

Proposition 3 *Let* $u_0, v_0 \in E_{(0,+\infty)}$, $u_0 \neq v_0$, *and satisfy*

(1) u_0 *is comparable with* v_0;

(2) the (H)difference $v_0 - u_0$ *exist;*

(3) u_0, v_0 *are on the same side of* $\tilde{1}$.

Then $\log_{\tilde{a}} u$ *is integrable over* $[u_0 v_0]$, *and the equality*

$$[\int_{u_0}^{v_0} F(u)\,du]^r = \begin{cases} [\underline{v_0}(r) - \underline{u_0}(r), \overline{v_0}(r) - \overline{u_0}(r)][\underline{w_0}(r), \overline{w_0}(r)] & \text{if } a > 1 \\ [\underline{v_0}(r) - \underline{u_0}(r), \overline{v_0}(r) - \overline{u_0}(r)][\overline{w_0}(r), \underline{w_0}(r)] & \text{if } 0 < a < 1 \end{cases}$$

holds for any $r \in [0,1]$, *where*

$$\underline{w_0}(r) = \begin{cases} \frac{1}{\underline{v_0}(r) - \underline{u_0}(r)} \log_a \frac{\underline{v_0}(r)^{\underline{v_0}(r)}}{\underline{u_0}(r)^{\underline{u_0}(r)}} - \frac{1}{\ln a} & \text{if } \underline{u_0}(r) \neq \underline{v_0}(r) \\ \log_a \underline{u_0}(r) & \text{if } \underline{u_0}(r) = \underline{v_0}(r) \end{cases}$$

$$\overline{w_0}(r) = \begin{cases} \frac{1}{\overline{v_0}(r)-\overline{u_0}(r)} \log_a \frac{\overline{v_0}(r)^{\overline{v_0}(r)}}{\overline{u_0}(r)^{\overline{u_0}(r)}} - \frac{1}{\ln a} & if\, \overline{u_0}(r) \neq \overline{v_0}(r) \\ \log_a \overline{u_0}(r) & if\, \overline{u_0}(r) = \overline{v_0}(r) \end{cases} .$$

Conclusions

Three results about the Zadeh extension (denoted by $\log_{\tilde{a}} u$) of function $\log_a x$ have been obtained. The first points out the continuity, monotonicity and convexity, and presents a representation of cut-form of $\log_{\tilde{a}} u$. the second gives out the directed G-differentiability of $\log_{\tilde{a}} u$, and presents a formula of calculating the directed G-derivative of $\log_{\tilde{a}} u$. The third gives out the integrablility (over a directed line segment) of $\log_{\tilde{a}} u$, and presents a formula of calculating the integration (over a directed line segment) of $\log_{\tilde{a}} u$.

Acknowledgments

This work is supported by a grant of Natural Science Foundation of China (10271035).

References

[1] S.S.L. Chang and L.A. Zadeh, "On fuzzy mapping and control", *IEEE Trans. Systems Man Cyberet.* **2** (1972) 30 34.

[2] P. Diamond and P. Kloeden, *Metric Spaces of Fuzzy Sets*, (World Scientific, 1994).

[3] D. Dubois and H. Prade, "Towards fuzzy differential calculus part 1: integration of fuzzy mappings", *J. Fuzzy sets and Systems* **8** (1982) 1 17.

[4] R. Goetschel and W. Voxman, "Elementary fuzzy calculus", *J. Fuzzy Sets and Systems* **18** (1986) 31 43.

[5] O. Kaleva, "Fuzzy differential equations", *J. Fuzzy Sets and Systems* **24** (1987) 301–317.

[6] S. Nanda and K. Kar, "Convex fuzzy mappings", *J. Fuzzy Sets and Systems* **48** (1992) 129–132.

[7] Wang Guixiang, "The G-differentiability and geometric properties of the Zadeh extensions of functions", Proc. The Third International Conference on Machine Learning and Cybernetics 2004, Shanghai, 1845 1849.

[8] Wang Guixiang and Wu Conxin, "Fuzzy n-cell numbers and the differential of fuzzy n-cell number value mappings", *J. Fuzzy Sets and Systems* **130** (2002) 367 381.

[9] Wang Guixiang and Wu Conxin, "Directional derivatives and subdifferential of convex fuzzy mappings and application in convex fuzzy programming", *J. Fuzzy Sets and Systems* **138** (2003) 559 591.

[10] Wang Guixiang and Wu Congxin, "The integral over a directed line segment of fuzzy mapping and its applications", *J. International Journal of Uncertainty, Fuzziness and Knowledge-Based Systems* **12** (2004) 543 556.

[11] Wu Conxin and Wang Guixiang, "Convergence of sequences of fuzzy numbers and fixed point theorems for increasing fuzzy mappings and application", *J. Fuzzy Sets and Systems* **130** (2002) 383 390.

Brill Academic Publishers
P.O. Box 9000, 2300 PA Leiden,
The Netherlands

Lecture Series on Computer and Computational Sciences
Volume 2, 2005, pp. 59-62

Numerical Simulation of Temperature Field 9SiCr Alloy Steel During Gas Quenching

Cheng Heming[1], Jin Min, Xie Jianbin

Department of Engineering Mechanics,
Faculty of Civil Engineering and Architectrue,
Kunming University of Science and Technology ,
650093 Kunming, China

Received 11 April, 2005; accepted in revised form 24 March, 2005

Abstract: The gas quenching is a modern, effective processing technology. On the basis of non-linear surface heat-transfer coefficient obtained by Ref. [1] during the gas quenching, the temperature field of 9SiCr alloy steel during gas quenching was simulated by means of finite element method. In the numerical calculation, the thermal physical properties were treated as the functions of temperature. The obtained results show that the non-linear effect should be considered in numerical simulation during gas quenching.

Keywords: gas quenching, surface heat-transfer coefficient, temperature, finite element method, 9SiCr alloy steel
Mathematics Subject Classification: Finite element methods, quench, temperature

PACS: 78M10. 33.50.Hv, 94.10.Dy

1 Introduction

The gas quenching is a modern, effective processing technology. Because of the smaller temperature difference of between the surface and middle of specimen in gas quenching, the residual stresses in specimen after gas quenching are smaller than that after water or oil quenching. This environment is polluted less with this technique, and it is prevailing in the industries[1,2,3,4]. But the researches on the mechanism during gas quenching are behind its applications. In order to obtain the distribution of residual stresses and perfect mechanical properties, it is necessary to control the phase transformation and limit distortion. Up to now thermal strains and thermal stresses can't be measured, the numerical simulation technology is an effective approach to understand the distribution and variety of thermal strains, thermal stresses and microstructure. In recent years, the numerical simulation of quenching processing in various quenching media is prevailing in the world. Although there are now some special soft-wares, which can simulate quenching processing, namely DEFORM-2, DEFORM-3, the important problem in numerical simulation is the boundary condition of stress and temperature. The calculating accuracy of thermal stresses and strains is closely related with the calculating precision of temperature field. For gas quenching processing, the key parameter of the calculation of temperature is the surface heat-transfer coefficient (SHTC).

On the basis of non-linear surface heat-transfer coefficient (SHTC) obtained by Ref. [1] during the gas quenching, the temperature field of 9SiCr alloy steel during gas quenching was simulated by means of finite element method. In the numerical calculation, the thermal physical properties were treated as the functions of temperature. The obtained results show that the inelastic-coupled effect should be considered during quenching.

2 Surface heat-transfer coefficients

During quenching, the surface heat-transfer coefficients (SHTC) have a great influence upon the microstructure and residual stresses in steel specimen. For this reason, many researches into this property have been studied. The variation of this property with temperature has been the subject of

[1] Corresponding author. E-mail: chenghm@public.km.yn.cn

investigations. The results obtained are very sensitive to small variation in the experimental conditions, which may lead to considerable discrepancies in the value obtained. Therefore, it must be found necessary to determine the effect of temperature on the surface heat transfer coefficient while using the actual experimental conditions that were to use during the subsequent determination of thermal stress and strain. R.F.Prince and A.J.Fletcher (1980)[5] have an effectual method, which can determine the relationship between temperature and surface heat transfer coefficient during quenching of steel plate. In the Ref.[1], the explicit finite difference method, nonlinear estimate method and the experimental relation between temperature and time during quenching have been used to solve the inverse problem of heat conduction. The relationships between surface temperature and surface heat transfer coefficient of cylinder have been given (as shown in Fig.1). Fig.2 denotes the comparison of surface heat-transfer coefficient in various quenching media.

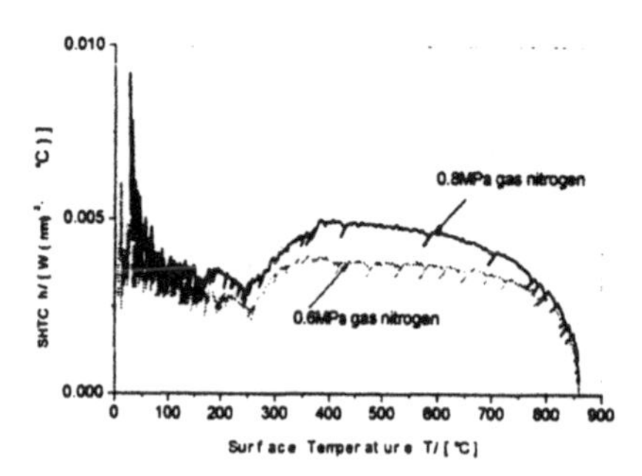

Figure 1 Surface heat-transfer coefficient during gas quenching

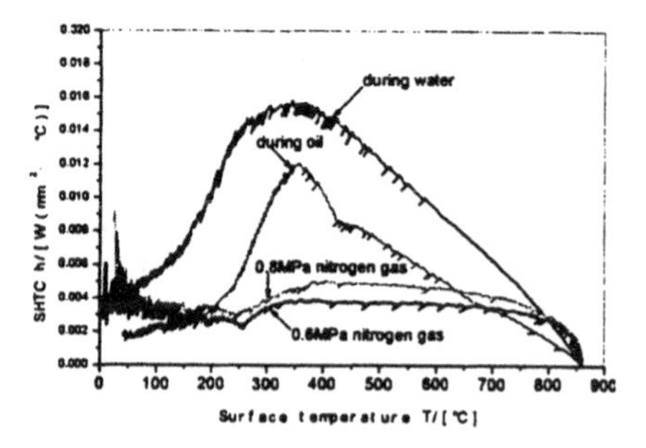

Figure 2 Comparison of surface heat-transfer coefficient between during various quenching media

The Figure 1 and Figure 2 show that (1). The surface heat-transfer coefficients appear as the stronger non-linear property for surface temperature. Therefore, it is inaccurate to simulate the processing of gas quenching by means of linear surface heat-transfer coefficients. (2). In the initial stage of gas quenching, there are stronger heat exchanges between specimen and gas.

3 The Functional and Finite Element Formula

A $\phi 20 \times 60mm$ cylinder of 9SiCr alloy steel was quenched by 0.8 MPa nitrogen gas from a temperature of 850 . The non-linear heat conduct equation is[6,7,8,9]

$$\lambda \frac{\partial^2 T}{\partial z^2} + \lambda \frac{\partial^2 T}{\partial r^2} + \lambda \frac{1}{r}\frac{\partial T}{\partial r} = \rho C_p \frac{\partial T}{\partial t} \tag{1}$$

where C_v and λ denote the specific heat capacity and the thermal conductivity, which are the function of temperature. Therefore, Eq. (1) is a nonlinear equation. The boundary condition of heat transfer is

$$\left.\frac{\partial T}{\partial n}\right|_{\Gamma} = h(T)(T_a - T_\infty) \tag{2}$$

The initial condition is

$$T\big|_{t=0} = T_0(x_k) \qquad k = 1,2,3 \tag{3}$$

here $h(T)$ is the surface heat-transfer coefficients, which are the functions of temperature and the volume fraction of phase constituents and was determined by finite difference method , non-linear estimate method and the experimental relationships between time and temperature[2]。 T_a and T_∞ denote the surface temperature of a cylinder of 45 steel and the temperature of quenching media. The functional[9] above problem is

$$K_n = -\int_{t_{n-1}}^{t_n}\int_\Omega \{\frac{\lambda}{2}[(\frac{\partial T}{\partial r})^2 + (\frac{\partial T}{\partial z})^2] + C_\rho T\dot{T}\}d\Omega dt +$$

$$+\int_{t_{n-1}}^{t_n}\int_\Gamma h(T_{n-1})\left(\frac{T^2}{2} - T_x T\right)dsdt \tag{4}$$

where λ and C_v, ρ denote the volume fraction of constituent, thermal conductivity and specific heat capacity, respectively. $h(T_{n-1})$ was the surface heat-transfer coefficients at t_{n-1}, $\{\dot{T}\}$ is fix in variational operation [9].The 8 nodes iso-parameter element was adopted, the finite element formula is:

$$[K]\{T\} + [M]\{\dot{T}\} = \{F\} \tag{5}$$

here *[K]* is a conductive matrix, *[M]* is a heat capacity matrix, *{F}* is heat supply duo to heat convection on boundary.

4 Calculated results

For 9SiCr alloy steel, the thermal physical properties with temperature are listed in Table 1.

Table 1、 Non-linear relationship of thermal physical properties with temperature[1]

Temperature(℃)	20	100	200	300	400	500	600	700	800	900
ρC_v $(10^{-3}Ws/mm^3K)$	2.9	2.75	2.6	3.8	3.45	3.6	3.2	3.0	2.64	2.5
Temperature(℃)	20	100	200	400	500	600	690	740	800	900
λ $(10^{-2}W/mmK)$	1.7	5.5	5.5	5.6	5.85	6.0	6.1	8.8	6.2	6.3

Figure 3 denotes the cooling curves of temperature during various quenching media. Fig.4 denotes the comparison of temperature field between the calculated values and experimental values.

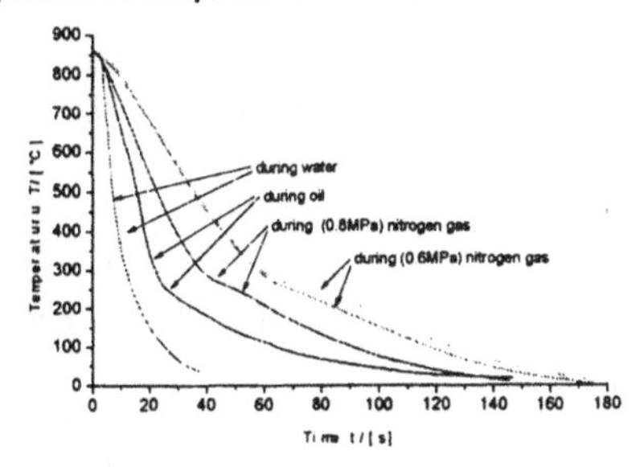

Figure. 3 Cooling curves of temperature during various quenching media

Figure.4 The comparison of temperature between calculated values and experimental values

5 Conclusions

(1)、 The heat conduction during gas quenching is a non-linear problem, the calculation of temperature is closely related with surface heat-transfer coefficient and phase transformation. For this reason, the effects of non-linear surface heat-transfer coefficient must be taken into account in simulation of temperature field during gas quenching.

(2) 、 Testing of cooling curves of work-pieces during nitrogen gas quenching process shows that the temperature difference of interior work-piece is small. It can be predicted that cooling of the interior work-piece is almost homogeneous and corresponding thermal strains and thermal stresses are rather small.

(3) 、 Different from water and spindle oil quenching, during gas quenching type of gas can be chosen and gas pressure can be adjusted, and the control of quenching parameters can be achieved by this way.

Acknowledgments

This work has been supported by the National Natural Science Foundation of China (10162002) and the Key Project of Chinese Ministry Education (No.204138)

References

[1] Xie Jianbin, *The numerical simulation and application of metal and alloy quenched in various quenching media.* Doctor's Dissertation, Kunming, Kunming University of Science and Technology, 2003.

[2] Holm T., Segerberg S., *Gas Quenching Branches out*, Advanced Materials and Processes, 149(6), 64W-64Z, 1996

[3] Segerberg S., *High-Pressure Gas Quenching in Cold Chamber for Increased Cooling Capacity,* 2M Int. Conf. on Quenching and Control of Distortion, Cleveland, Ohio, USA, 4-7 Nov.,1996.

[4] Midea S.J., Holm T., Segerberg S., High-Pressure Gas Quenching-Technical and Economical Consideration, 2M Int. Conf. on Quenching and Control of Distortion, Cleveland, Ohio, USA, 4-7 Nov., 1996.

[5] R.F.Prince and A.J. Fletcher, *Determination of surface heat-transfer coefficients during quenching of steel plates.* J. Met. Tech., **2:** 203-215.1980

[6] Cheng Heming, He Tianchun, Xie Jianbin, Solution of an inverse problem of heat conduction of 45 steel with martensite phase transformation in high pressure during gas quenching, *Journal of materials science and technology*, **18(4)**:372-374. 2002

[7] Cheng Heming, Wang Honggang, Chen Tieli, Solution of an inverse problem of heat conduction of 45 steel during quenching, *Acta Metallurgica Sinica*, **33** (**5**) : 467 1997 (in Chinese)

[8] J. A. Adams and D. F. Rogers, Computer-aided heat transfer analysis, 1973, McGraw-HillC.

[9] Wang Honggang, Theory of Thermo-elasticity, Press Qinghua University,1989 (in Chinese)

Brill Academic Publishers
P.O. Box 9000, 2300 PA Leiden,
The Netherlands

Lecture Series on Computer and Computational Sciences
Volume 2, 2005, pp. 63-66

Monte Carlo Method for Simulation of a Laser Beam Propagating in a Cavity with Specular and Diffuse Components of Reflection

C.Y. Ho[1]

Department of Mechanical Engineering,
Hwa Hsia Institute of Technology,
235, Taipei, Taiwan, Republic of China

Received 8 April, 2005; accepted in revised form 20 April, 2005

Abstract: A laser beam incident on a cavity wall with specular and diffuse components of reflection is partially absorbed, and the remainder diffusely and specularly reflected by the wall. A Monte Carlo method is utilized to simulate the process of propagation of a laser beam in the cavity and calculate the radiant energy absorbed by the cavity wall. In this study the intensity of a laser beam is assumed to be TEM_{00} mode of a Gaussian distribution and the cavity is chosen to be a paraboloid of revolution. The way dividing the cavity wall into grid is presented for the numerical calculation. The integrated energy calculated by a random generator with a cycle of 2^{31}-1 is compared with that with a cycle of 2^{20}. The convergence was tested by changing different numbers of energy bundles, ring-elements, and subdivisions. The results reveal that the energy flux absorbed by the cavity abruptly rises near the critical radius for the cavity depth enough to make the incident beam intervene more than two specular reflections in the cavity.

Keywords: Monte Carlo method, Laser beam, Specular reflection, Diffuse reflection, Paraboloid of cavity

PACS: 05.10.Ln, 52.50.Jm, 44.40.+a, 42.68.Ay, 41.85.-p

1. Assumptions and analysis

A Monte Carlo method is used to study the radiant energy absorbed by a paraboloid of revolution-shaped cavity subject to a laser-beam. The incident flux is partially absorbed, and the remainder diffusely and specularly reflected by the cavity wall. The surface on the cavity has specular and diffuse components of reflection. Therefore, Reflections of irradiation are composed of specular and diffuse components. The assumptions made for Monte Carlo method and physical model are listed follows:

1. The geometric optics theory is valid for this analysis of radiant heat transfer in the welding or drilling cavity [1].
2. The various effects and quantities are additive and the directional change of a pencil of radiation is negligible. The former assumption implies that the phenomena of diffraction, interference and coherence are excluded [1]
3. Reflections of irradiation are composed of specular and diffuse components [2-4].
4. The cavity is idealized by a paraboloid of revolution to a first approximation [5,6]. The cavity is stationary even though drilling and welding are unsteady.
5. Planck's and Kirchhoff's laws are valid.

2. Dimensionless incident flux and apparent absorptivity

With the above assumptions, the incident flux of a TEM_{00} mode can be represented by Gaussian distributions

$$q = \frac{3}{\pi\sigma^2} exp[-3(\frac{r}{\sigma})^2] \tag{1}$$

where constant 3 is chosen such that 95 percent of incident energy is included within the energy-distribution radiusσ.
Apparent absorptivity of the cavity is defined as

$$\alpha_a \equiv \frac{\text{Total absorbed energy in cavity}}{\text{Total incoming energy or beam power}} \equiv Q_a(1) \tag{2}$$

[1] Corresponding author. E-mail: hcy2126@cc.hwh.edu.tw

where $Q_a = Q_a(r)$ represents integrated energy absorbed by the cavity within a circular area of radius r. Evidently, apparent absorptivity is identical to the dimensionless integrated absorption at a dimensionless radius of 1.

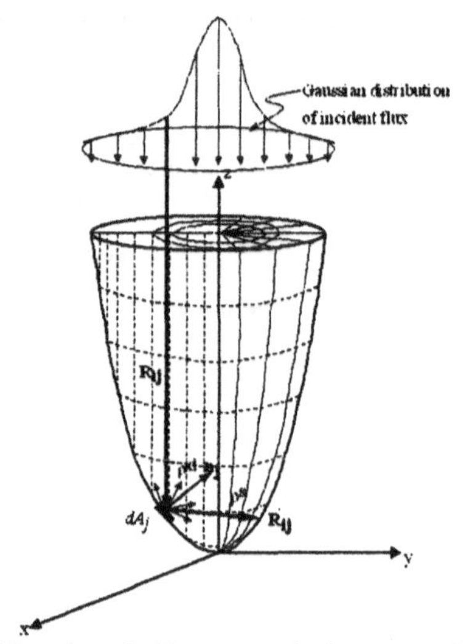

Figure 1: Schematic of laser beam incident on a cavity in a rectangular coordinate system

3. Monte Carlo method

3.1 Specular reflection component

If the incident vector between the receiving point j and starting point i on the cavity is known, the specularly reflected ray after striking point j can be described by the law of coplanarity and the law of reflection.

$$\mathbf{R}_{ij}^{'} = \frac{-2\mathbf{n}_j(\mathbf{n}_j \bullet \mathbf{R}_{ij}) + \mathbf{R}_{ij}}{\left| -2\mathbf{n}_j(\mathbf{n}_j \bullet \mathbf{R}_{ij}) + \mathbf{R}_{ij} \right|} \quad (3)$$

Hence the specularly reflected vector $\mathbf{R}_{ij}^{'}$ of the incident energy-beam is accomplished [7,8].

3.2 Diffuse reflection component

If N_i rays are imagined to leave dA_i, all with energy Q_i^d / N_i, where Q_i^d is the diffuse energy radiated from surface dA_i. The sum of energies of the rays absorbed by surface dA_j therefore yields $\varepsilon_i \, Q_i^d \lim_{N_i \to \infty} (N_{ij} / N_i)$ [2,4,9,10]. The quantity N_{ij} represents the number of energy bundles emitted by surface dA_i, which are eventually absorbed at surface dA_j. The next step is to find the probability that an energy bundle will be emitted or reflected from the surface in a given direction. It is convenient to express the direction of bundle emission or reflection in terms of the cone angle θ and polar angle φ of a spherical coordinate system centered at the emitting or reflected location. The probability function is normalized, so that it takes values between zero and unity. For a diffuse surface, the probability distribution functions of bundle emission or reflection for the cone and polar angles, respectively, yield

$$P_\varphi = \frac{\varphi}{2\pi}, \qquad P_\theta = \sin^2\theta \quad (4)$$

Once the probability functions are established, the path of the energy bundle can be traced. The direction (θ, φ) is found by generating a pair of random numbers P_θ, P_φ from a uniformly distributed set between zero and unity. Suppose that the surface location on which the bundle is incident has an absorptivity α. A random number P_φ is drawn from a uniformly distributed set between zero and one. If $0 \le P_\varphi \le \alpha$, the incident bundle is absorbed, the absorption is tallied and attention is redirected to the next energy bundle leaving the point of emission. On the other hand, the bundle is reflected for $\alpha \le P_\varphi \le 1$.

A uniform distribution and independent random sequence of random numbers are required to determine the directions of the diffuse component in a Monte Carlo method. Hence, the speed of generation and cycle are of considerable importance. The random numbers generated by arithmetical processes are not in a random fashion but pass some criterion of randomness. The choice of a random generator depends on the particular problem and computer used. In this work, a random generator with a cycle of 2^{31}-1 is used and compared with that with cycle of 2^{20}. Convergence has been tested by changing different numbers of energy bundles, ring-elements, and subdivisions. Choosing numbers of energy bundles N_b = 1000-3000, ring elements N_r = 100-300, and subdivisions N_e = 10-30, a relative deviation of the absorbed energy flux was generally found to be less than 1 percent.

4. Results and discussion

Figure 2 shows that the variations of the integrated energy absorbed by the cavity with radial distance for different random generators and diffuse reflectivities. The solid line represents the integrated energy calculated by a random generator with a cycle of 2^{31}-1.On the other hand the dashed line indicates the integrated energy calculated by a random generator with a cycle of 2^{20}. For the case of $\rho^d = 0.0001$ the integrated energies absorbed by the cavity wall are almost equal for these two random generators with different cycles and the relative errors increase with the increasing diffuse reflectivity. This is because the random generator plays an important role in diffuse reflections. The cycle of random generator evidently influences the calculated results, special in the location of radial coordinate around r=0.6.

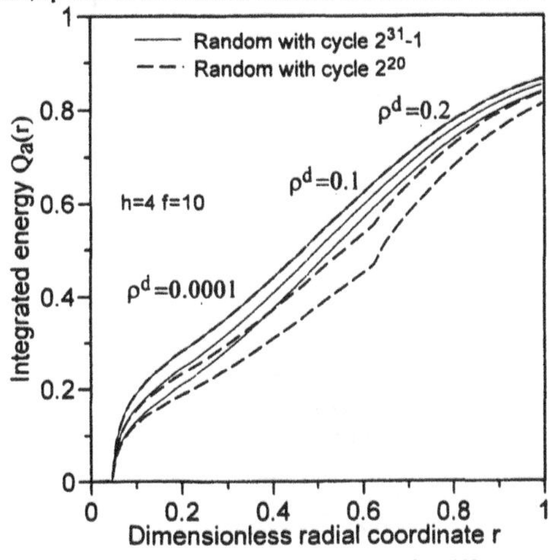

Figure 2: Integrated energy absorbed by the cavity for different random generators.

Convergences were tested by changing numbers of energy bundles, numbers of ring-elements, cavity depths and reflectivites. The apparent absorptivities for different numbers of energy bundles and ring-elements, cavity depths and reflectivites were listed on table 1. The relative error was found to be less than 1 percent.

Table 1: Apparent absorptivities for different numbers of ring-elements and energy bundles

Apparent absorptivity α_a		α=0.5, pure diffuse		α=0.5, ρ^s=0.3, ρ^d=0.2	
Number of ring-elements	Numbers of energy bundles	h=1	h=10	h=1	h=10
150	1000	0.67327	0.98469	0.69508	0.88630
150	2000	0.67286	0.98470	0.69511	0.88627
150	3000	0.67291	0.98469	0.69514	0.88626
100	2000	0.67285	0.98470	0.69435	0.88624
300	2000	0.67290	0.98470	0.69440	0.88626

Dimensionless energy flux absorbed and integrated incident energy and absorption versus radius at different dimensionless cavity depths are shown in Fig. 3. The dimensionless energy flux absorbed is referred to the absorbed energy per unit area of the cavity wall. The integrated incident and absorbed energies at a given radial location are calculated by integrating energy fluxes from the axisymmetric axis to this location. The energy flux absorbed by the cavity of h=3 is low near r=0, and then abruptly increase when r approaches $1/4h^2$ (=1/36). After that the energy flux rapidly decreases with the increasing radial distance away from the cavity base again. However the profile of the energy flux absorbed by the cavity of h=0.5 is similar to the Gaussian distribution of incident laser beam. For the case of h=3, the region $r < 1/4h^2$ only receives the directly incident energy and diffusely reflected energy from other locations but the region $r > 1/4h^2$ receives multiple specular reflections and diffuse reflections from other locations besides the directly incident energy. Therefore the sudden jump for the energy flux absorbed occurs at around $r=1/4h^2$. The critical radius $r=1/4h^2$ is derived from geometric optics for a paraboloid of revolution-shaped cavity. For a shallow cavity h=0.5, the incoming rays escape from the cavity after only intervening one specular reflection so the energy absorbed by the cavity approximately equals to the product of absorptivity and the Gaussian-profiled incidence flux. The integrated incident energy and absorption increase with the increasing radial distance from symmetric axis. It is also worth to note that the integrated absorption increases more quickly near the region of the rise of the energy flux for the case h=3.

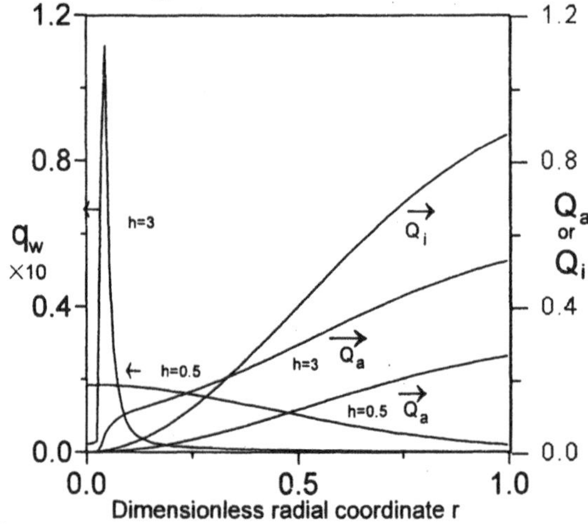

Figure 3: Energy flux and integrated energy absorbed by the cavity wall.

Acknowledgments

The author wishes to thank the support for this work by National Science Council of the Republic of China under grant no. NSC 93-2212-E-146-001-.

References

[1] M. Born, and E. Wolf: *Principles of Optics: Electromagnetic Theory of Propagation, Interference and Diffraction of Light*. 7th edition, Cambridge University Press, UK, Chapters 1 and 3, 1999.
[2] K. E. Torrance, and E. M. Sparrow, *J. Opt. Soc. Am.* **57** 1105-1114(1967).
[3] K. E. Torrance, *J. Heat Transfer* **91** 287-290(1969).
[4] B. van Ginneken, M. Stavridi, and J. J. Koenderink, *Appl. Optics* **37** 130-139(1998).
[5] H. Tong, and W. H. Giedt, *The Review of Scientific Instruments* **40** 1283-1285(1969).
[6] P. S. Wei, T. H. Wu, and Y. T. Chow, *J. Heat Transfer* **112** 163-169(1990).
[7] M. F. Modest: *Radiative Heat Transfer*. McGraw-Hill, New York, Chapters 2-6, 1993.
[8] S. C. Wang, and P. S. Wei, *Metall. Trans. B* **23B** 505-511(1992).
[9] J. S. Toor, and R. Viskanta, *Int. J. Heat Mass Transfer* **11** 883-897(1968).
[10] R. Siegel, and J. R. Howell: *Thermal Radiation Heat Transfer*. 3rd edition, Taylor and Francis, Washington D. C., Chapters 4 and 5, 1992.

Brill Academic Publishers
P.O. Box 9000, 2300 PA Leiden,
The Netherlands

Lecture Series on Computer and Computational Sciences
Volume 2, 2005, pp. 67-70

Numerical Simulation of Temperature Field for T10 Steel during High Pressure Gas Quenching

Lijun Hou [1]; Heming Cheng, Jianyun Li

Department of Engineering Mechanics,
Kunming University of Science and Technology,
Kunming 650093, P.R. China

Received 11 April, 2005; accepted in revised form 21 April, 2005

Abstract: The explicit finite difference method, nonlinear estimate method and the experimental relation between temperature and time during gas quenching have been used to solve the inverse problem of heat conduction. The relations between surface heat-transfer coefficient in 6×10^5 Pa pressure and surface temperature of steel cylinders are given. Based on the solved surface heat-transfer coefficients, the temperature field is obtained by solving heat conduction. In numerical calculation, the thermal physical properties of material are treated as the functions of temperature. The results of calculation coincide with the results of experiment.

Keywords: finite difference method; temperature field; surface heat-transfer coefficient; gas quenching

1. Introduction

It is well-known that the mechanical properties can be improved by quenching technique. High pressure gas quenching is a new heat treatment technique. The temperature field of workpiece has a great influence upon the residual stresses and microstructure after quenching. Therefore, the calculation of temperature field during quenching has been the subject of many investigations.

In this paper, a T10 steel cylinder was taken as an investigating example. Above all, the non-linear relations between surface heat-transfer coefficient and surface temperature of steel cylinder were obtained by solving the inverse problem of heat conduction. Finally, the temperature field was calculated based on the surface heat-transfer coefficients dependent of temperature.

2. Heat Conduction Equation and Finite Difference Method

The specimen is a ϕ 20mm×60mm cylinder of T10 steel. The heat conduction equation in middle of specimen is

$$C_V\rho\frac{\partial T}{\partial t}=\lambda\left(\frac{\partial^2 T}{\partial r^2}+\frac{1}{r}\frac{\partial T}{\partial r}\right) \tag{1}$$

the boundary condition is

$$\lambda\left.\frac{\partial T}{\partial r}\right|_{\Gamma}=h(T_S-T_\infty)+C_V\rho\frac{\partial T}{\partial t} \tag{2}$$

[1] Corresponding author. E-mail: houlijun978@yahoo.com.cn

where λ, C_V and ρ denote thermal conductivity, specific heat capacity and density of material respectively. C_V and λ are the functions of temperature. T, r and t denote temperature, radius and time. h, T_S and T_∞ denote surface heat-transfer coefficient, specimen temperature of surface boundary and temperature of quenching media, respectively.

The finite difference equations of Eq. (2) and Eq. (1) are

$$\lambda \frac{T_i^2 - T_i^1}{\Delta r}\bigg|_\Gamma = h(T_i^1 - T_\infty) + C_V\rho\frac{T_i^1 - T_{i-1}^1}{2\Delta t} \tag{3}$$

$$T_{i+1}^j = [1 - \frac{\Delta t}{(C_V\rho)_i^j \Delta r}(\frac{\lambda_i^{j+1} + \lambda_i^j}{2\Delta r} + \frac{\lambda_i^j + \lambda_i^{j-1}}{2r_j})] \times T_i^j + [\frac{\Delta t(\lambda_i^{j+1} + \lambda_i^j)}{2(C_V\rho)_i^j \Delta r}(\frac{1}{\Delta r} - \frac{1}{r_j})] \times T_i^{j+1}$$
$$+ [\frac{\Delta t(\lambda_i^j + \lambda_i^{j-1})}{2(C_V\rho)_i^j \Delta r}(\frac{1}{\Delta r} + \frac{1}{r_j})] \times T_i^{j-1} \tag{4}$$

where λ_i^j and $(C_V\rho)_i^j$ denote thermal conductivity and specific heat capacity at j node in i time step. Δr, r_j and Δt denote node interval, node radius and time step, respectively.

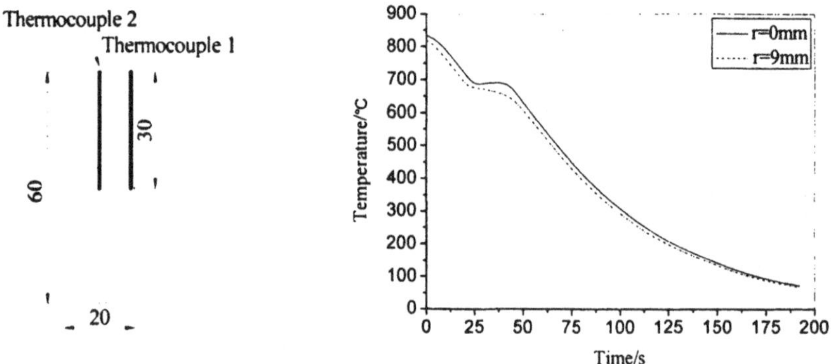

Figure 1: Specimen and position of thermocouple. Figure 2: Experimental values of temperature.

3. Experiment

The change of temperature during quenching was measured by thermocouples. Experimental data were recorded by a computer. The thermocouple 1 is placed 1mm distance from the cylinder surface of specimen, thermocouple 2 is on the axis of the cylinder, as shown in Fig. 1. The temperature values measured by thermocouple 1 are key parameters for solving the nonlinear heat conduction equation. The temperature value's measured by thermocouple 2 were used to examine the calculated results. The specimen was heated up to 820 , and austenized for 20 min. Then, it was quenched by nitrogen with 6×10^5 Pa pressure. The measured values of temperature are shown in Fig. 2.

4. Calculation Method of Surface Heat-transfer Coefficient and temperature field during Quenching

Nonlinear Estimate Method

In order to solve the inverse problem of this nonlinear heat conduction equation, the nonlinear estimate method is adopted. Let a parameter

$$\varphi = \sum (T^{\exp} - T_{i+1}^{cal})^2 \tag{5}$$

to be minimized. We have

$$\frac{\partial \varphi}{\partial h} = (T^{\exp} - T_{l+1}^{cal})\frac{\partial T_{l+1}^{cal}}{\partial h} = 0 \tag{6}$$

where $T^{\exp}$ is the measured temperature, T_{l+1}^{cal} is ($l+1$)th iterative computations value of temperature. T_{l+1}^{cal} can be written as Taylor series

$$T_{l+1}^{cal} = T_l^{cal} + \frac{\partial T}{\partial h}\Delta h \tag{7}$$

With the help of Eq. (6), Eq. (7) and the convection condition on boundary, we obtain an incremental surface heat-transfer coefficient Δh, that is

$$\Delta h = \frac{\lambda}{T_S - T_\infty}\frac{T_l^{cal} - T^{\exp}}{\Delta r} \tag{8}$$

Table 1: The thermal properties of T10 steel in different temperatures.

Temperature/	20	200	300	400	500	700	900
$\lambda/10^{-2}$(W/(mm·K))	4.154	4.320	4.238	4.082	3.890	3.549	3.524
$C_V\cdot\rho/10^{-3}$(W·s/(mm^3·K))	3.284	4.031	4.146	4.273	4.435	4.791	6.663

Calculating Procedure

The thermal properties ($\lambda, C_V\cdot\rho$) at various node were calculated by using the linear interpolation method according to Table 1. After an estimate value h_0 of surface heat-transfer coefficient was given, the heat conduction equation (4) can be solved and the computation temperature values at node are obtained. If $|T_l^{cal} - T^{\exp}| > \delta$, a new value of h ($h = h_0 + \Delta h$) will be chosen. T_{l+1}^{cal} is repeated until $|T_l^{cal} - T^{\exp}| \le \delta$, in which δ is an inherent error. In this way, the changes of surface heat-transfer coefficient in temperature and the temperature field are obtained. In the calculation, time interval is automatically obtained. According to the stability of explicit finite difference scheme, the maximum Δt is chosen as the time interval during calculation.

$$\Delta t \le \frac{2\Delta r C_V \rho}{\frac{\lambda_{j+1} + 2\lambda_j + \lambda_{j-1}}{\Delta r} - \frac{\lambda_{j+1} - \lambda_{j-1}}{r_j}} \tag{9}$$

5. The Calculated results

The relations between surface heat-transfer coefficient and surface temperature are non-linear during high pressure gas quenching, as shown in figure 3. In the initial period during quenching, the increase of the surface heat-transfer coefficient is rapid. This shows that the exchange of the heat flux between the workpiece and quenching media is large. It is noted that at temperature of 700 , there is a sudden change in the surface heat-transfer coefficient, because of pearlite phase transformation in this temperature. After the end of pearlite phase transformation, the change in the surface heat-transfer coefficient is small.

The calculated result of temperature field is shown dash dot line in figure 4. These five curves denote the change of temperature with time at the position of $r = 0, 2, 4, 6, 8mm$ of the cylinder, respectively. In the calculation of temperature field, the non-linear relations between surface heat-transfer coefficient and surface temperature were taken into account. At the axis (at $r = 0$), the results of calculation coincide well with the results of experiment.

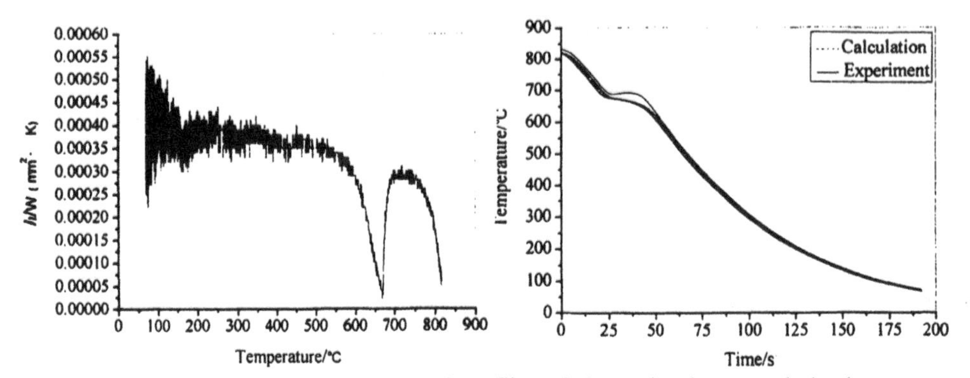

Figure 3: Relations between surface heat-transfer coefficient and surface temperature.

Figure 4: Comparison between calculated temperature field and experimental values of temperature.

Acknowledgements

This project has been supported by the National Natural Science Foundation of China (10162002) and the Key Project of Chinese Ministry of Education (No. 204138).

References

[1] CHENG Heming, WANG Honggang and CHEN Tieli, Solution of Heat Conduction Inverse Problem for Steel 45 during Quenching, *ACTA METALLURGI SINICA*, 33(5), 467-472(1997) (in Chinese).

[2] Heming CHENG, Tianchun HE and Jianbin XIE, Solution of an Inverse Problem of Heat Conduction of 45 Steel with Martensite Phase Transformation in High Pressure during Gas Quenching, *J. Mater. Sci. Technol.*, 18(4), 372-374(2002).

[3] Hu Mingjuan, Pan Jiansheng, Li Bing and Tian Dong, Computer Simulation of Three Dimensional Temperature Field of Quenching Process with Suddenly Changed Surface Conditions, *TRANSACTIONS OF METAL HEAT TREATMENT*, 17(Suppl.), 90-97(1996) (in Chinese).

[4] Beichen Liu, *Engineering Computation Mechanics--Theories and Application.* China Machine Press, Beijing, 1994 (in Chinese).

[5] Heming Cheng, Jianbin Xie and Jianyun Li, Determination of Surface Heat-transfer Coefficient of Steel Cylinder with Phase Transformation during Gas Quenching with High Pressures, *Computational Materials Science*, 29(4), 453-458(2004).

[6] XIE Jianbin, CHENG Heming, HE Tianchun and WEN Hongguang, Numerical Simulation of Temperature Field in Steel 1045 during Its Quenching, *Journal of Gansu University of Technology*, 29(4), 33-37(2003) (in Chinese).

[7] Z. Li, R.V. Grandhi and R. Shivpuri, Optimum Design of the Heat-transfer Coefficient during Gas Quenching Using the Response Surface Method, *International Journal of Machine Tools & Manufacture*, 42(5), 549–558(2002).

Brill Academic Publishers
P.O. Box 9000, 2300 PA Leiden,
The Netherlands

Lecture Series on Computer and Computational Sciences
Volume 2, 2005, pp. 71-74

Numerical Simulation of Turbulent Flow through a Water-Turbine Spiral Casing

Huying[1] Cheng Heming[2] Li Ziliang[3]

1.School of Mechanical and Electrical Engineering,
Kunming University of Science and Technology ,Yunnan, P.R. of China
2.School of Civil Engineering and Architecture,
Kunming University of Science and Technology, Yunnan, P.R. of China
3. .School of Mechanical and Electrical Engineering,
Kunming University of Science and Technology ,Yunnan, P.R. of China

Received 10 April, 2005; accepted in revised form 24 April, 2005

Abstract: Based on the Navier-Stokes equations and the standard k-ε turbulence model, the mathematical model for the turbulent flow through a turbine spiral casing is set up when the boundary conditions, including inlet boundary conditions, outlet boundary conditions and wall boundary conditions, have been implemented. The computation and analysis on internal flow through a spiral casing have been carried out by using the SIMPLEC algorithm and CFX-TASCflow software for fluid flow computation so as to obtain the simulating flow fields. The calculation results at the design operating condition for the spiral casing are presented in this paper. Thereby, an effective method for computing the internal flow field in a spiral casing has been explored .

Keywords: Water-Turbine Spiral Casing, Turbulent Flow, Numerical Simulation

Mathematics Subject Classification: 76F99

PACS: 47.27.Eq

1. Introduction

The hydraulic design methods of a turbine spiral casing include the equal velocity momentum method($v_u r=c$),the equal circular velocity method($v_u=c$), the variable velocity momentum method($v_u r=k(\varphi)$), the section area specific method and the variable average circumferential velocity method. As some hypotheses are made in these methods, the hydraulic performance of the spiral casing which is designed by them is not enough ideal. Along with the rapid development of computational techniques and computational fluid dynamics, it has been an important method to optimize the spiral design by predicting the performance of a spiral casing by means of numerical simulation.

2. Mathematical Model

Numerical simulation is based on the mathematical model to be set up. For Newtonian fluid, it is the Navier-Stokes equations and boundary conditions.

[1] Corresponding author. E-mail: huying2003@vip.km169.net

[2] Corresponding author. E-mail: chenghm@public.km.yn.cn

[3] Corresponding author. E-mail: lziliang@public.km.yn.cn

2.1. Governing Equations

Time-averaging of N-S equations, the Reynold equations for turbulent flow may be expressed as:

$$\frac{\partial \bar{\rho}}{\partial t}+\frac{\partial}{\partial x_j}(\bar{\rho}\bar{u}_i)=0 \tag{1}$$

$$\frac{\partial}{\partial t}(\bar{\rho}\bar{u}_i)+\frac{\partial}{\partial x_j}(\bar{\rho}\bar{u}_j\bar{u}_i)=-\frac{\partial \bar{p}}{\partial x_i}+\frac{\partial}{\partial x_j}(\bar{\tau}_{ij}+\overline{\rho u_i' u_j'})+\rho\bar{u}_i \tag{2}$$

In the above equations u_i and u_j represents the velocities in the coordinate directions, p is the static pressure, τ_{ij} is the viscous stress tensor. For Newtonian fluids,

$$\tau_{ij}=\mu(\frac{\partial u_i}{\partial x_j}+\frac{\partial u_j}{\partial x_i})-\frac{2}{3}\mu\frac{\partial u_l}{\partial x_l}\delta_{ij}$$

When a fluid is modeled as incompressible, the Reynolds stress and turbulent flux terms are related to the mean flow variables using an eddy viscosity assumption.

$$\overline{\rho u_i' u_j'}=\rho\overline{u_i' u_j'}=\mu_t(\frac{\partial \bar{u}_i}{\partial x_j}+\frac{\partial \bar{u}_j}{\partial x_i})-\frac{2}{3}\rho\delta_{ij}k \tag{3}$$

Where, μ_t is an eddy viscosity coefficient. Equation (3) can only express the turbulent fluctuation term as functions of the mean variables if the turbulent kinetic energy k and turbulent viscosity μ_t are known. The standard k- ε turbulence model provides these variables[3].

$$\mu_t=\rho C_u\frac{k^2}{\varepsilon} \tag{4}$$

$$\frac{\partial(\rho k)}{\partial t}+\frac{\partial(\rho\bar{u}_j k)}{\partial x_j}=\frac{\partial}{\partial x_j}(\Gamma_k\frac{\partial k}{\partial x_j})+P_k-\rho\varepsilon \tag{5}$$

$$\frac{\partial(\rho\varepsilon)}{\partial t}+\frac{\partial(\rho\bar{u}_j\varepsilon)}{\partial x_j}=\frac{\partial}{\partial x_j}(\Gamma_\varepsilon\frac{\partial \varepsilon}{\partial x_j})+\frac{\varepsilon}{k}(C_{\varepsilon_1}P_k-\rho C_{\varepsilon_2}\varepsilon) \tag{6}$$

Where, $\Gamma_k=\mu+\frac{\mu_t}{\sigma_k}$, $\Gamma_\varepsilon=\mu+\frac{\mu_t}{\sigma_\varepsilon}$, $P_k=\mu_t(\frac{\partial \bar{u}_i}{\partial x_j}+\frac{\partial \bar{u}_j}{\partial x_i})\cdot\frac{\partial \bar{u}_i}{\partial x_j}$

$C_u=0.09$, $C_{\varepsilon 1}=1.44$, $C_{\varepsilon 2}=1.92$, $\sigma_k=1.0$, $\sigma_\varepsilon=1.3$[4]

The closed non-linear equation set which can describe the flow in a spiral casing consists of the above equations (1), (2), (3), (4), (5) and (6).

2.2. Boundary Conditions

After the mathematical model is set up, it is important to specify boundary conditions so as to solve the equations. So called boundary conditions is the law that the solved variables vary with coordinates on the boundary faces. It's impossible to get the solutions of a flow field until the most appropriate set of boundary conditions are chosen.

The following boundary conditions on the turbulent flow through a spiral casing are presented in this paper.

2.2.1 Inlet Boundary Conditions

An inlet boundary condition is one where the fluid enters the domain, i.e. the inlet of a spiral casing. It is usually supposed the fluid flow is well-distributed at an inlet. Thus, the inlet velocity can be required by computations or tests. It is permitted that the inlet velocity be specified as:

$$\Phi=\Phi_{spec}\ (\Phi=u,v,w)$$

2.2.2 Outlet Boundary Conditions

An outlet boundary condition is one where the fluid exits the computational domain, i.e. the outlet of a spiral casing. At the outlet, the flow is approximately fully developed and doesn't vary along flow directions. A surface perpendicular to the flow direction is chosen as an outlet surface. Then, outlet boundary conditions can be implemented. The mathematical description of outlet boundary conditions are defined by

$$\frac{\partial \phi}{\partial n} = 0, \quad (\varphi = u,\ v,\ w,\ k,\ \varepsilon)$$

Where n is locally normal to the outlet boundary face.

2.2.3 Wall Boundary Conditions

A no-slip wall exists where a viscous flow is in contact with a solid object. Right at the wall the fluid is stationary with respect to the solid object. Then, the fluid velocity normal to the wall is zero, i.e.

$$u=v=w=0$$

3 The Results and Analysis of Numerical Simulations

The analysis and computation on flows through the spiral casing of a model hydraulic turbine applied for Sanbanxi hydropower are performed by the CFX-TASC flow software for fluid flow. The results of numerical simulation of turbulent flows are presented in this paper. It is a metal round section spiral casing and is designed by the equal velocity momentum method ($V_u r=k$) .Its entrance diameter is $\Phi 448.5mm$. The calculation operating condition unit flow is $Q_{11}=0.56m^3/s$.

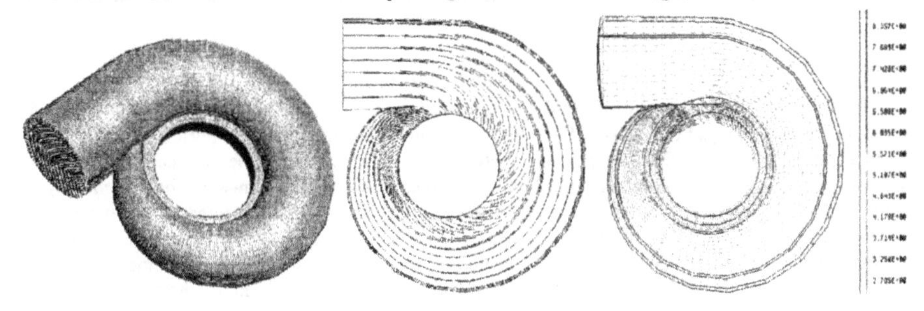

Figure 1: Entire computational grid for the spiral casing

Figure 2: Simulated flows through the spiral casing

Figure3:velocity on the symmetrical plane

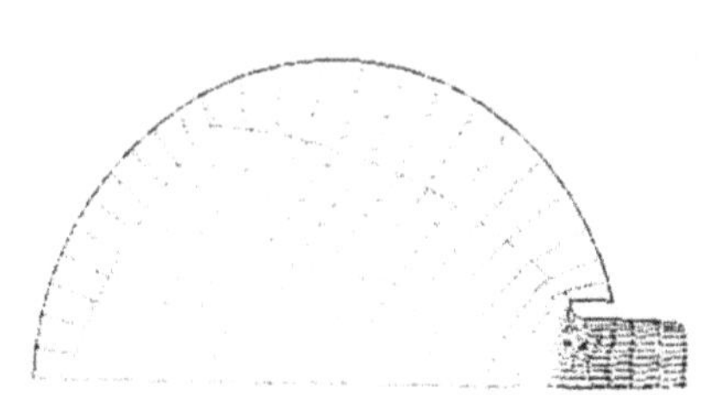

Figure4: Projection of the velocity field at the wrapped angle $\varphi=0°$ cross-section

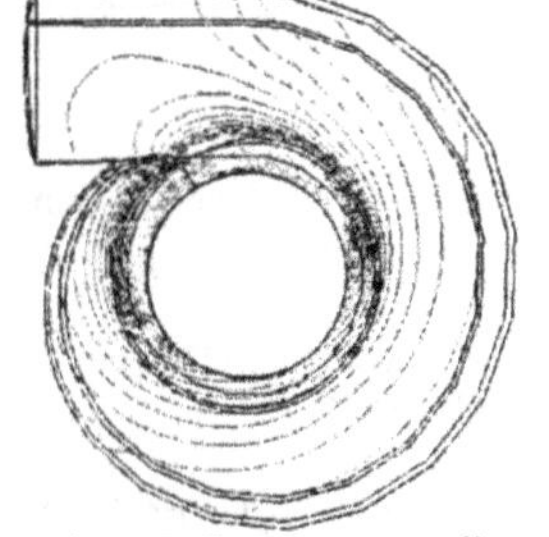

Figure5: Equal pressure lines in the spiral casing

Fig.1 illustrates the entire computational grid for the spiral casing. $11\times70\times120=92400$ grid nodes are used. Fig.2 displays the simulated flows through the spiral casing. Fig.3 shows the velocity on the symmetrical plane. The projection of the velocity field at the wrapped angle $\varphi=0°$ cross-section is shown in Fig.4. The numerical results of the velocity field in the spiral casing show that the flow velocity is well-distributed, the distribution of streamlines and the velocity vectors are quite well too.

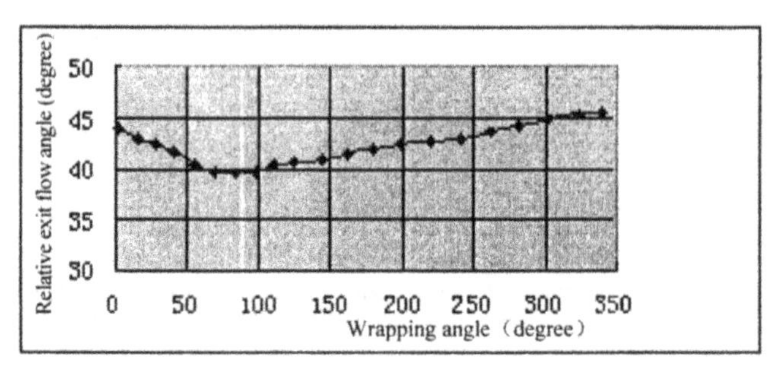

Figure 6: variation of relative flow angle at the exit of a spiral casing along the wrapping angle

Fig.5 represents the equal pressure lines in the spiral casing. It shows the pressure is well-distributed. The relative flow angles around the circle (the circumscribed circle of stay vanes) are calculated so as to quantitate the flow variation along the circle. The variation of relative flow angle at the exit of a spiral casing along the wrapping angle are shown in Fig.6. The difference of the maximal and minimal relative flow angle is 7° or so. It should be a line parallel to the axial x theoretically. But it is impossible to do so, as the size of a spiral casing is limited and the flow is three-dimensional one. Therefore, the fluid flow into equably along the circular as far as possible, so as to reduce energy losses in a spiral casing and improve performance of a hydraulic turbine. The numerical results show that the performance of the spiral casing is high.

4. Conclusions

Based on N-S equations, the standard k- ε turbulence model and the boundary conditions, after the computation and analysis on internal flow through a spiral casing have been performed by applying TASCflow, the simulating flow fields at the sections of a spiral casing can be obtained. With the help of flow analysis for the spiral casing, spiral casing profiles of high performance can be designed by the modification and improvement of an existing spiral casing. The results presented in this paper show that numerical flow simulations can be integrated successfully into a design procedure for a spiral casing by CFD method, which can ensure performance of it, dramatically shorten the time needed for a new one, reduce numbers of test and design cost. It is an available and efficient method of design.

5. Acknowledgments

This work was supported by National Science Foundation of China(No.10162002) and Key Project of Chinese Ministry Education (No.204138).

References

[1] Mingde Su, Foundation of Computational Fluid Dynamics, Tsinghua University Press, Beijing, 1997 (in Chinese).

[2] TASCflow Theory Documentation, Advanced Scientific Computing Ltd., Waterloo, Ontario, Canada, 1995.

[3] Zhongqing Jin, Numerical Solution of NS Equations and Turbulent Model, Hehai University Press, Nanjing, 1989 (in Chinese).

[4] B.E.Launder and D. B. Spalding, The Numerical Computation of Turbulent Flows, Computer Methods in Applied Mechanics and Engineering, 3 (2), 269-289(1974).

[5] Fujun Wang, Analysis of Computational Fluid Dynamics, Tsinghua University Press, Beijing, 2004 (in Chinese).

[6] Merle C. Potter, David C. Wiggert, Mechanics of Fluids, China Machine Press,Beijing,2003.

Brill Academic Publishers
P.O. Box 9000, 2300 PA Leiden,
The Netherlands

Lecture Series on Computer and Computational Sciences
Volume 2, 2005, pp. 75 - 78

Dynamic Delegation using Privilege Storage in Centralized Administration

Jun-Cheol Jeon, Kee-Won Kim and Kee-Young Yoo[1]

Department of Computer Engineering,
Kyungpook National University,
702-701 Daegu, Korea

Received 9 April, 2005; accepted in revised form 24 April, 2005

Abstract: Role-based access control (RBAC) has been recognized as an efficient access control model for secure computer systems. In the meantime, centralized administration has received a lot of attention due to a simple and strong control mechanism in access control. However, it is really hard to know what controls are appropriate for every object/subject when the number of object/subject is very large by a single authority. Thus delegation of authority could be delayed or uncontrollable so that it is not suitable for real time systems. In this paper, we propose a dynamic delegation strategy in role based delegation model for flexible and automatic management. To realize delegation mechanism, we provide an efficient method for separation of duty rule and privilege distribution. Thus, it blocks that user acting alone can compromise the security of the data processing system, and it also minimizes frequent granting operations and overburden to certain users for continuous works.

Keywords: Centralized Administration, Role-based Access Control, Delegation and Revocation, Separation of Duty, Right and Responsibility.

Mathematics Subject Classification: Dynamical system in control

PACS: 37N35

1. Motivation of research

Access control is arguably the most fundamental and most pervasive security mechanism in use today. Access control shows up in virtually all systems and imposes great architectural and administrative challenges at all levels of enterprise computing. From a business perspective, access control has the potential to promote the optimal sharing and exchange of resources, but it also has the potential to frustrate users, impose large administrative costs, and cause the unauthorized disclosure or corruption of valuable information [1]. In the mean while, RBAC models have matured to the point where they are now being prescribed as a generalized approach to access control. For instance, recently RBAC was found to be "the most attractive solution for providing security features in multi domain digital government infrastructure." [2], and has shown its relevance in meeting the complex needs of Web-based applications [3].

One of the key issues in most access control systems is concerned with authorization administration policy, which refers to the function of granting and revoking authorizations. Centralized and decentralized administrations are two possible approaches to policy management [4]. With centralized administration, it is rather inflexible, since usually no individual can know what controls are appropriate for every object/subject when the number of object/subject is very large. Meanwhile, in

[1] Corresponding author. E-mail: yook@knu.ac.kr, jcjeon33@hotmail.com

decentralized administration, it is flexible and apt to the particular requirements of individual subjects. Nevertheless, the authorizations become more difficult to control since multiple subjects can grant and revoke authorizations, and the problem of cascading and cyclic authorization may arise.

The administrative functions can be roughly divided into two parts: 1) creation and deletion, and 2) monitoring and maintenance [1]. The authorization for creation and deletion could be simply realized according to the policy demanded by the given environment. However, the authorization for monitoring and maintenance is relatively hard to control which objects and subjects are suitably assigned when a number of delegations are occurred for a number of objects and subjects.

To make simple and robust administration environment, we propose the strategy in delegation model within centralized administration. Privilege generally consists of right and responsibility, it means that privilege always accompanies with right and responsibility. Therefore, for stable and robust delegation mechanism, two functions, separation of duty (SoD) rule and privilege distribution will be considered. SoD is dividing right for critical operations so that no user acting alone can compromise the security of the data processing system, and well-designed distribution of privilege can minimize frequent granting operations and dissatisfaction of users.

2. Dynamic delegation strategy using privilege storage

Figure 1 shows an example of the proposed dynamic architecture in which each role, $Ri(1 \leq i \leq 5)$, has a *storage*, except for *R6* since the users assigned in bottom role can not delegate any privileges. The architecture is composed of *layers* in role hierarchies, the roles toward the top of the hierarchy represent the more powerful roles (i.e. those roles containing a greater number of authorized privileges, including rights and responsibilities, and fewer authorized users), and the roles toward the bottom of the graph represent the more general roles (i.e. those roles containing fewer authorized privileges and a greater number of authorized users).

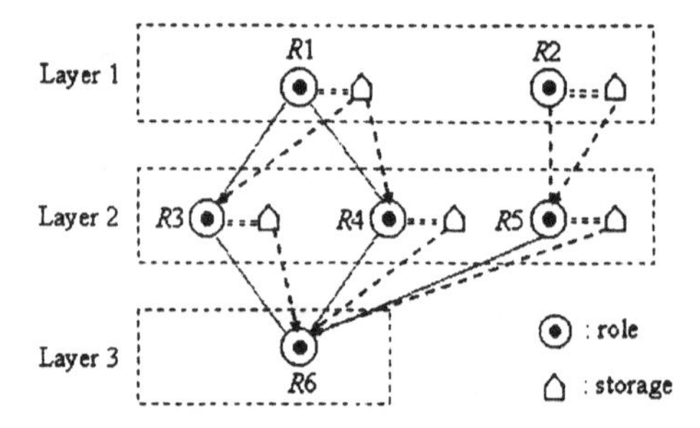

Figure 1: Example architecture of proposed model: The solid lines indicate the role hierarchy and the dotted arrow lines indicate delegable privilege to role assignments, and the double dotted lines indicate delegable privilege to storage assignments.

We suppose that every layer has equivalent responsibility, and the members assigned a role also have the same responsibility represented as *score*. Each privilege has a *label* corresponding to its gravity and a sort of work: *T* and *N* for trivial and nontrivial operations, respectively, and we use a numeral for describing a sort of privilege: e.g. a privilege has a label of *T1* then it shows that the privilege executes a trivial operation and its sort is '1'. That is, the privilege with label '1' cannot be delegated to a user who already has a privilege with label '1'.

All user information are recorded in database system including *current score*, *charged score*, *hierarchy information*, *timestamp* when a privilege is deposited into a storage, *sort number* which he/she has, and *layer number* where he/she is located. Consequently, our algorithm finds oldest privilege first by timestamp, and finds smallest scored user, then match them. For the following example, suppose that the members in layer 1, 2, and 3 have a responsibility of 20, 15, and 10 points respectively.

Suppose Alice is assigned to *R*1, and John, Tom, and Sahra are assigned to *R*3, *R*4, and *R*5, respectively shown in Table 1. John, Tom and Sahra are working with current score 10, 12 and 12 points respectively. At this moment, Alice wants to delegate a privilege, (*settle_count*, *T*1). We may decide the capacity of the privilege corresponding to the label such that *T* and *N* are regarded as 1 point and 2 points respectively. Then, we find the best suitable grantee so that Tom would be selected as a proper grantee even though John is the least scored member in Layer 2, because John is working with sort number, '1'.

Table 1. Example of current scores, charged scores, and sort of current privileges corresponding roles and users.

Role	*R*1	*R*2		*R*3	*R*4	*R*5	*R*6
User	Alice	Bob	Carol	John	Tom	Sahra	Chris
Current score	19	18	19	10	12	12	9
Charged score	20	20		15	15	15	10
Sort #	1,2	3	2	1	2	4	5
Layer #	1			2			3

Thus we block that many privileges are centralized in a particular user using the label, *T* or *N*, and the constraint of SoD can be preserved by checking sort numbers. It is optional that Sahra may be chosen as the suitable grantee, if the system does care a role hierarchy or not.

By the way in the real world, we cannot ignore the admission sequence of privileges into the storage. It means that we need to consider not only the gravity of privilege but also the sequence of admission for deciding candidates. Suppose that there are three privilege suites, (*handle_mail*, *T*1), (*issue_check*, *N*2), and (*settle_count*, *N*2) at a storage. Then we may take *issue_check* or *settle_count* first as candidates even if *handle_mail* came into the storage earlier. It makes a trouble with mail handle (i.e. opens mail, reviews, and endorses checks, etc.) while the privilege is staying at the storage. Thus the time information is added to the privilege suite when a privilege is assigned to storage. The following algorithm describes an efficient discriminating algorithm for finding suitable grantees.

```
            [Delegation Algorithm: Search_and_Match algorithm]

             Input : (privilege, label, timestamp)
             Output: (privilege, grantee)
             Initialize: the value of T, N, and Charged Score

While there exists a suite (privilege, label, timestamp) at the storage

Label 1:  Choose an oldest privilege suite;
                Check the timestamp
          Check the layer's number;
          Check the sort of privilege;
          Call search_and_match procedure; ·

End

Procedure search_and_match
   The number of layer + 1;
   Find the least scored member;
         If the member are performing a privilege with the sort number;
         Then go to the Label 1;
         Else go to the next;
   Available Score =: Charged Score - Current Score
         If the Available Score is bigger than the value of the
   privilege, T or N;
         Then go to the next stage;
         Else go to the Label 1;
   If the member is in role hierarchy (optional);
   Then match it with the privilege.
   Else eliminate it;
End
```

First of all, we choose a delegable privilege at a storage based on first-come-first-service by the timestamp. Secondly, we find a best suitable grantee by label and scores. Lastly we match the privilege to proper grantee. It might be thought that a grantor would like to delegate a specific privilege to the particular member. It is possibly achieved for flexible management by administrator's acceptance.

In revocation mechanism, we bring our focus into the term of delegation such as temporal delegations and permanent delegations. Temporal delegation refers to the type of delegation that is limited by time or intention of the grantor. One that time is expired, the delegation is no longer valid, and can be revoked the delegation as grantor wishes. We apply the revocation using time out for temporal delegation. It can be represented as a privilege suite using duration time or expired time, i.e. (*privilege, label, timestamp, duration time*) or (*privilege, label, timestamp, expired time*). In the mentioned above example, privilege suite can be represented as (*handle_mail, T1, 5-Jan-2004(16:40), 3 Days*) or (*settle_count, N2, 5-Jan-2004(16:40), 8-Jan-2004(16:40)*).

3. Discussion and Conclusion

We have proposed an efficient delegation mechanism based on the policies of SoD requirement and granting responsibility within central administration circumstance. We have taken advantage of a storage in which delegable privileges are deposited. Thus the following merits have been obtained:

- It has worked automatically without anyone's interference so that it can block the malicious attention and is suitable for real time systems.
- We have observed the constraint of separation of duty by classifying a sort of privileges.
- Efficient sharing responsibility by well-defined policy has minimized frequent granting operations and dissatisfaction of users.
- We have possibly controlled the amount of privileges by dividing trivial and nontrivial operations.

In addition to that, our scheme has provided much flexibilities that system administrator can decide to initialize the number of a sort of works, the maximum score assigned to user in each layer, and the score associated with trivial or nontrivial operations, etc. These characteristics help to use the proposed mechanism more generally.

In this paper, we have described a new notion of dynamic privilege delegation model. All privileges have been transferred to storage for delegation, and the privileges at the storage have been automatically assigned to the best suitable grantee immediately by proposed algorithm. We have applied privilege suites that each of suites consists of a privilege, a label, a timestamp, a specified grantee, and duration or expired time. Our scheme has not required the supervision of system administrator and it is controlled automatically during monitoring and maintaining. Thus our mechanism can be effectively used for the granting strategy in complex delegation systems within centralized administration environments.

Acknowledgments

The authors would like to thank the anonymous referees for their valuable suggestions and fruitful comments on how to improve the quality of the manuscript. This work was supported by the Brain Korea 21 Project in 2005.

References

[1] David F. Ferraiolo, D. Richard Kuhn, and Ramaswamy Chandramouli, *Role-Based Access Control*, Artech house, 2003

[2] James B. D. Joshi, Arif Ghafoor, Walid G. Aref, and Eugene H. Spafford, Digital Government Security Infrastructure Design Challenges, *IEEE Computer*, 2(34), pp. 66-72 (2001).

[3] James B. D. Joshi, Walid G. Aref, Arif Ghafoor, and Eugene H. Spafford, Security Models for Web-Based Applications, *Communication of the ACM*, 2(44), pp. 38-72 (2001).

[4] Chun Ruan and Vijay Varadharajan, A formal graph based framework for supporting authorization delegations and conflict resolutions, *International Journal of Information Security*, 1(4), pp. 211-222 (2003).

Brill Academic Publishers
P.O. Box 9000, 2300 PA Leiden,
The Netherlands

Lecture Series on Computer and Computational Sciences
Volume 2, 2005, pp. 79-82

Efficient and Secure User Identification Scheme based on ID-based Cryptosystem[1]

Kee-Won Kim, Jun-Cheol Jeon, Kee-Young Yoo[2]

Department of Computer Engineering, Kyungpook National University, Daegu, KOREA, 702-701

Received 10 April, 2005; accepted in revised form 25 April, 2005

Abstract: Tseng–Jan proposed ID-based cryptographic schemes based on Maurer–Yacobi's scheme. Hwang–Lo–Lin modified Tseng–Jan's user identification scheme to make it suitable for the wireless environment. In this paper, we show that Hwang–Lo–Lin's user identification scheme is vulnerable to the forgery attack. Therefore, we propose an improvement of Hwang–Lo–Lin's scheme to resist to the forgery attack. The proposed scheme not only eliminates the security leak of Hwang–Lo–Lin's scheme, but also reduces communication and computation costs as compared with their scheme.

Keywords: Identification, ID-based Cryptosystem

1 Introduction

In 1984, Shamir proposed the idea of a public key cryptosystem in which the public key can be an identification information [1]. In this cryptosystem, the public key of the user is verified without using the services of a third party. Each user uses a unique identification information (e.g., name, address and e-mail address) as his public key. Under Shamir's idea, many identity-based public cryptosystems have been proposed [2, 3]. In 1991, Maurer–Yacobi developed a identity-based non-interactive public key distribution system based on a novel trapdoor one-way function of exponentiation modulo a composite number [4]. However, some problems with this scheme were found, the scheme was modified [5] and the final version was presented [6]. In 1998, Tseng–Jan improved the scheme proposed by Maurer–Yacobi, and provided a identity-based non-interactive public-key distribution system with multi-objectives such as an identity-based signature scheme, an identification scheme, and a conference key distribution system [7]. However, in the wireless environment, the capacity of the battery of a mobile device is limited. The time for waiting and responding must be reduced. Therefore, Hwang–Lo–Lin proposed an efficient user identification scheme based on Tseng–Jan's scheme that is suitable for the wireless environment [8].

In this paper, we point out that Hwang–Lo–Lin's user identification scheme is vulnerable to the forgery attack. Therefore, we propose an improvement of Hwang–Lo–Lin's scheme to resist to the forgery attack. The proposed scheme outperforms Hwang–Lo–Lin's scheme in both aspects of the communication and the computation costs. The structure of our paper is organized as follows. We briefly review Hwang–Lo–Lin's user identification scheme in Section 2. In Section 3, we show that Hwang–Lo–Lin's scheme is vulnerable to the forgery attack. In Section 4, we propose the improved scheme to enhance the security. In Section 5, we analyze the security of our improved

[1]This work was supported by the Brain Korea 21 Project in 2005.
[2]Corresponding author. E-mail: yook@knu.ac.kr

scheme. In Section 6, the performance of the proposed scheme is discussed. Finally, we give a brief conclusion in Section 7.

2 Review of Hwang–Lo–Lin's scheme

Hwang Lo Lin proposed an efficient user identification scheme based on Tseng Jan's scheme. We briefly review Hwang Lo Lin's user identification scheme in this section. Hwang–Lo–Lin's user identification scheme consists of three phases: the initiation phase, the user registration phase and the user identification phase.

The initiation phase: For the system setup, a trusted authority (TA) chooses four primes p_j between 60 and 70 decimal digits where for each p_j such that the numbers $(p_j - 1)/2$ are odd and pairwise relatively prime, where $1 \leq j \leq 4$. Thus, let $N = p_1 \cdot p_2 \cdot p_3 \cdot p_4$. TA also selects an integer e in $Z^*_{\phi(N)}$ and computes the secret value d which satisfies $e \cdot d \equiv 1(\bmod \phi(N))$, where ϕ is the Euler's totient function. TA chooses a secret multiplier t at random from $Z^*_{\phi(N)}$ and computes $v \equiv t^{-1}(\bmod \phi(N))$.

The user registration phase: When a user U_m with the identity ID_m wants to join the system, TA computes the secret key of U_m as follows: $s_m \equiv e \cdot t \cdot \log_g(ID_m^2)(\bmod \phi(N))$, where g is a primitive element in $GF(p_j)$, for $1 \leq j \leq 4$. TA sends s_m to U_m as his secret key in secret. Let $h(\cdot)$ be a one-way function. TA publishes $\{N, g, e, h(\cdot)\}$ and keeps $\{p_1, p_2, p_3, p_4, t, v, d\}$ secret for all users. The identity ID_m of U_m is the public key of U_m and s_m is the secret key of U_m.

The user identification phase: The mobile device (M) wants to show his identity ID_m is legal to the base station (BS). The identity of BS is ID_b. Hwang Lo Lin's user identification scheme is presented as follows.

Step 1. Mobile device (M) chooses a random integer k in Z^*_N and computes Y and Z as follows: $Y = (ID_m^2)^k \bmod N$ and $Z = (ID_b^2)^{k \cdot s_m \cdot T} \bmod N$, where T is the current time used as a timestamp. Then, he sends $I = \{(ID_m||Y||Z), T\}$ to the base station (BS).

Step 2. After receiving the message I from M, BS computes $Z' = Y^{s_b \cdot T} \bmod N$.

Step 3. BS checks the equation $Z \stackrel{?}{=} Z'$. If equivalent, BS will confirm that M's identity is valid.

3 Weakness of Hwang–Lo–Lin's user identification scheme

In this section, we show that Hwang Lo Lin's user identification scheme is vulnerable to the forgery attack. The adversary intercepts the message $I = \{(ID_m||Y||Z), T\}$ from network during the legal user M sends it to the BS. The adversary computes Y' and Z'' as follows: $Y' = (Y)^T \bmod N$ and $Z'' = (Z)^{T'} \bmod N$, where T' denotes the adversary's identification time used a new timestamp. Hence, the forged message is constructed as follows: $I' = \{(ID_m||Y'||Z''), T'\}$. Thus the forged message I' will satisfy verification equation in Step 3. This is because

$$\begin{aligned} Z' &= (Y')^{s_b \cdot T'} \bmod N \qquad (1) \\ &= (Y)^{T \cdot s_b \cdot T'} \bmod N \\ &= (ID_m^2)^{k \cdot T \cdot s_b \cdot T'} \bmod N \\ &= (g^{s_m \cdot d \cdot v})^{k \cdot T \cdot s_b \cdot T'} \bmod N \\ &= (g^{s_b \cdot d \cdot v})^{k \cdot T \cdot s_m \cdot T'} \bmod N \\ &= (ID_b^2)^{k \cdot s_m \cdot T \cdot T'} \bmod N \\ &= Z^{T'} \bmod N \\ &= Z''. \end{aligned}$$

4 The proposed scheme

In this section, we propose an improvement on Hwang Lo Lin's schemes to resist the forgery attack and then analyze the proposed scheme. The initiation phase and the user registration phase are the same as those of Hwang Lo Lin's scheme. The mobile device (M) wants to show his identity ID_m is legal to the base station (BS). The identity of BS is ID_b. The proposed scheme is presented as follows.

Step 1. Mobile device (M) computes Z as follows: $Z = (ID_b^2)^{s_m \cdot T} \bmod N$, where T is the current time used as a timestamp. Then, he sends $I = \{(ID_m||Z), T\}$ to the base station (BS).

Step 2. After receiving the message I from M at the time T^*, BS checks whether $(T^* - T)$ is within the valid time interval ΔT. If not, the identification message is rejected.

Step 3. BS computes $Z' = (ID_m^2)^{s_b \cdot T} \bmod N$ and checks whether the following equation holds: $Z \stackrel{?}{=} Z'$. If equivalent. BS will confirm that M's identity ID_m is valid.

5 Security analysis

The security of the proposed scheme is based on the same cryptographic assumptions as those in Hwang Lo Lin's scheme. In the following, some attacks are presented.
Attack 1: An adversary tries to reveal M's secret key s_m from $Z = (ID_b^2)^{s_m \cdot T} \bmod N$ of the message I. It is equivalent to the difficulty of computing the discrete logarithm modulo the composite number N. Thus, M's secret key s_m will never be revealed to the public. Similarly, it is difficult to reveal BS's secret key s_b from Z of the message I.
Attack 2: Suppose that an adversary records $I = \{(ID_m||Z); T\}$ to cheat BS that he/she is the legal user M. If the adversary tries to replay $I = \{(ID_m||Z), T\}$ to BS, BS would reject it because the adversary cannot pass the verification $(T^* - T) \leq \Delta T$ in Step 2. If the adversary replaces T into a new timestamp T' and sends $I' = \{(ID_m||Z), T'\}$ to BS, BS would reject it because the adversary cannot pass the verification $Z \stackrel{?}{=} (ID_m^2)^{s_b \cdot T'}$ in Step 3. To replace a new timestamp T' in Z requires solving the difficult factoring problem.
Attack 3: An adversary tries to forge a valid identification message $I = \{(ID_m|| Z), T\}$. Since the adversary has no knowledge of M's secret key s_m, the adversary cannot compute the valid $Z = (ID_b^2)^{s_m \cdot T} \bmod N$. Thus, the adversary cannot forge the valid identification message.
Attack 4: An adversary tries to mount the forgery attack in Section 3 on the proposed scheme. The adversary computes $Z'' = (Z)^{T'} \bmod N$ with Z and a new timestamp T' and sends $I' = \{(ID_m||Z''), T'\}$ to BS. BS would reject it because the adversary cannot pass the verification $Z'' \stackrel{?}{=} (ID_m^2)^{s_b \cdot T'}$ in Step 3. Thus, the adversary cannot mount the forgery attack in Section 3.

6 Performance analysis

In this section, we compare the proposed scheme and Hwang Lo Lin's scheme in terms of computation costs and communication costs. The notation T_{exp} denotes the time for performing a modular exponentiation computation and $|x|$ denotes the bit-length of x. The comparisons between the proposed scheme and Hwang Lo Lin's scheme are stated in Table 1. It can be seen that the proposed scheme is more efficient than Hwang Lo Lin's scheme in both the computation costs and the communication costs. Therefore, the proposed scheme is suitable to be used in wireless environment.

Table 1: The comparison of Hwang Lo Lin's scheme and the proposed scheme

	Hwang Lo Lin's scheme	The proposed scheme
Communication costs	$\|ID\| + \|Y\| + \|Z\| + \|T\|$	$\|ID\| + \|Z\| + \|T\|$
Computation costs for M	$2T_{exp}$	$1T_{exp}$
Computation costs for BS	$1T_{exp}$	$1T_{exp}$

7 Conclusion

In this paper, we pointed out that Hwang Lo Lin's user identification scheme is insecure. An adversary can forge a user identification message. Moreover, this paper also proposed an improved user identification scheme to resist the forgery attack. Our improved scheme not only retains the merits of Hwang Lo Lin's scheme but also withstands the forgery attack. Moreover, the proposed scheme outperforms Hwang Lo Lin's scheme in both aspects of communication costs and the computation complexities. Therefore, the proposed scheme is suitable to be used in wireless environment.

References

[1] A. Shamir, Identity based cryptosystems and signature schemes, *Advances in Cryptology - Crypto'84*, Lecture Notes in Computer Science, Volume 196, Springer, 47-53(1984).

[2] S. Tsujii and T. Itoh, An ID-based cryptosystem based on the discrete logarithm problem, *Journal of Selected Areas in Communications*, **7**(4), 467-473(1989).

[3] D. Boneh and M. Franklin, Identity based encryption from the Weil pairing, *Advances in Cryptology - Crypto'01*, Lecture Notes in Computer Science, Volume 2139, Springer, 213-229(2001).

[4] U.M. Maurer and Y. Yacobi, Non-interactive public key cryptography, *Advances in Cryptology - Eurocrypt'91*, Lecture Notes in Computer Science, Volume 547, Springer, 498-507(1991).

[5] C.H. Lim and P.J. Lee, Modified Maurer-Yacobi's scheme and its application, *Advances in Cryptology - Asiacrypt'92*, Lecture Notes in Computer Science, Volume 718, Springer, 308-323(1992).

[6] U.M. Maurer and Y. Yacobi, A non-interactive public-key distribution system, *Designs, Codes and Cryptography*, **9**(3), 305-316(1996).

[7] Y.M. Tseng and J.K. Jan, ID-based cyprtographic schemes using a non-interactive public-key distibution system, *Proc. of the 14th Annual Computer Security Applications Conference(IEEE ACSAC98)*, 237-243(1998).

[8] M. S. Hwang, J.W. Lo and S.C. Lin, An efficient user identification scheme based on ID-based cryptosystem, *Computer Standards & Interfaces*, **26**(6), 565-569(2004).

Brill Academic Publishers
P.O. Box 9000, 2300 PA Leiden,
The Netherlands

Lecture Series on Computer and Computational Sciences
Volume 2, 2005, pp. 83-87

Evaluation of Success Factors in Mobile Commerce Using the AHP Techniques

Jongwan Kim[1], Chong-Sun Hwang[2]

Department of Computer Science and Engineering,
Korea University, Anam-Dong, Sungbuk-Ku,
136-701, Seoul, Korea

Received 12 April, 2005; accepted in revised form 26 April, 2005

Abstract: Commerce using mobile connections to the Internet is fast developing and the electronic commerce paradigm is shifting towards this mobility. Mobile commerce has some new aspects, mobility, that differ from e-commerce, and these differences increase the complexity of business factors. Success will probably be driven by a number of key business factors (B2C, B2B, B2E, and P2P). In this paper we analyze and rate the factors, with alternatives, for successful mobile commerce, using the Analytic Hierarchy Process. We surveyed the relative importance of factors and alternatives by directing a questionnaire to business managers, students and other users who are using mobile Internet, mobile banking and the other mobile services. The results show that the main factors that influence customers are related to trust and security. A structured analysis of such factors provides good insights, and will help business managers to time the launch of mobile commerce businesses.

Keywords: Analytic Hierarchy Process, Mobile commerce, Multiple Criteria Decision Making

1. Introduction

The advance in information technology from wire-connected Internet to mobile Internet access is radically affecting customer needs and purchasing patterns. In general, the decision-making model followed by a customer engaged in purchasing activity consists of five major phases, characterized by several components and involving one or more alternative actions. The five stages are needs identification, information search, evaluation of alternatives, purchase & delivery, and after-purchase evaluation [1]. The customer will use information from outside sources to develop a set of criteria and evaluate alternatives. Identification of these criteria and alternatives is an important consideration in preparing to launch a mobile commerce business, from the view point of a company. This study aimed elucidate the factors that affect success in mobile commerce, and then evaluate and rate these factors by analyzing components of commercial activity in the mobile Internet environment, using the Analytic Hierarchy Process (AHP).

Priorities are determined by the decision maker's preferences. These preferences are modeled by multi-criteria methods. Many research studies have used Multiple Criteria Decision Making (MCDM) [2] such as Multi Attribute Utility Theory (MAUT) [3], Analytical Hierarchy Process (AHP) [4]. AHP is a decision-aiding method that was developed by Saaty [4]. It is a multi-criteria decision method that utilizes structured pair-wise comparisons among systems of similar alternative strategies to produce a scale of preference.

Success in electronic commerce is an issue that has attracted the attention of many specialists [5]. The electronic commerce paradigm is shifting, and more participants are being attracted from e-commerce to wireless mobile commerce, but the decision alternatives for mobile commerce are, to some extent, still the same as in e-commerce.

---

[1] Ph.D. student and works for DISYS laboratory (distributed systems). E-mail: wany@disys.korea.ac.kr
[2] Professor and active member of IEEE. E-mail: hwang@disys.korea.ac.kr

In this paper we analyze and rate in order the factors and choices involved in successful mobile commerce. It is important to note that structured factors provide instructive insights that can help business managers to evaluate the undertaking of new mobile commerce business.

2. Criteria Affecting Commerce with Mobile Access

The term m-commerce covers an emerging set of applications and services that people can access from their Web-enabled mobile devices [6] using the "wireless Web". M-commerce inherits many attributes from e-commerce, and we have employed some e-commerce characteristics from the E-commerce Success Model [5].

In the mobile Internet environment, people can use a mobile application with a wireless connection anywhere and at anytime. Mobility of devices and applications raises the issue of the appropriateness of their use under certain circumstances [7], that is, mobility is a strategic consideration for m-commerce to utilize in aiming for success. We can find the electronic commerce success factors in [5] and extract the major aspects of m-commerce from [7]. These consist of seven m-commerce success factors: System Quality, Content Quality, Use, Trust, Support and Mobility, and alternatives for decision-making are shown below. More detailed information on these categories can be found in references [5], [6], [7] and [8].

- System Quality: Recent works focusing on e-commerce have suggested additional variables: *online response time, 24-hour availability page loading speed* and *visual appearance*. Since these considerations also apply to e-commerce, these criteria and their components are equally applicable to an m-commerce system [9], [10].
- Content Quality: Information systems' literature has emphasized the importance of information quality as one of the determinants of user satisfaction, and has identified a number of attributes: *up-to-dateness, understandability, timeliness* and *preciseness* [11].
- Use: The extent to which a system is used is one of the measures that are widely used to define the success of a business. Considering the purposes of e-commerce systems suggested by [12], the use of an e-commerce system can be divided into informational and transactional components. Such attributes are applied in exactly the same way to m-commerce. The informational use logged by a customer can be described as requesting and obtaining information. These terms are often shortened to *information* and *transaction*.
- Trust: Trust is another significant challenge in the m-commerce environment. Customers are concerned about the level of security when providing sensitive information online [13]. Also, they expect that personal information will be protected from external access. There are two alternatives – *security* and *privacy*.
- Support: Support is a customer-oriented criterion and includes the following components: *tracking order status, account maintenance, payment alternatives* and *FAQs* [14].
- Mobility: Mobility of *device* and *application* raises the issue of their suitability for the user under some circumstances [7].

3. Evaluation of the Success Factors through AHP

The Analytic Hierarchy Process is an MCDM technique to assist the solution of complex criteria problems. This has been found to be an effective approach that can settle complex decisions; it has been used by numerous researchers in various fields [15], [16] to handle both tangible and intangible factors and sub-factors. As the AHP approach is a subjective methodology, information and the priority weights of elements may be obtained from a decision-maker using direct questioning or by a questionnaire method.

Zahedi [17] summarized the AHP procedure in terms of four steps: (1) Break the decision problem into a hierarchy of interrelated problems. (2) Derive the matrix data for pair-wise comparison of the decision elements. (3) Find the eigenvalues and eigenvectors of the pair-wise comparison matrices to estimate the relative weights of the decision elements. (4) Aggregate the relative weights of the decision elements to obtain a rating for decision alternatives.

The AHP model was formulated and data were collected to assess the professional judgment of customers or decision-making executives using the "1-9 Saaty Scale" for pair-wise comparisons. The criteria and decision alternatives that are applied in this evaluation were described in Section 2. The results of this application provide good analytical penetration regarding m-commerce success factors in the market. The process of calculation is next described in more detail.

- *Break down Decision Problems (Step 1)*: In this step, we build a decision hierarchy by breaking a general problem into individual criteria. The top of the hierarchy is the overall objective, the decision alternatives are at the bottom. The middle nodes are the relevant attributes (criteria) of the decision problem. The number before each element designates the node number.
- *Pair-wise Comparison (Step 2)*: Next, we gather rational data for the decision criteria and alternatives, using the AHP relational scale suggested by Saaty. The results of this stage are presented as pair-wise comparison matrices of the decision elements indicating the global objective, criteria, and decision alternatives.
- *Normalize the Matrix (Step 3)*: we use the eigenvalue method to estimate the relative weights of the decision elements.
- *Estimate the Relative Priorities (Step 4)*: we perform a composition of priorities for the criteria that give the rank of each alternative.

4. Findings

At this stage we have calculated the pair-wise values and priorities (or weights) of alternatives. This provides a good basis for making decisions and selecting between alternatives, but the AHP may contain internal inconsistencies; the consistency ratio (CR) offers a method of obtaining rational assessment of the alternatives. It will assume the value zero in the case of perfect consistency and will be positive otherwise. If the results are inconsistent, after checking the consistencies, we reassess the assigned matrix values in an iterative manner until satisfactory consistency is achieved.

Table 1 shows that the criteria trust and mobility have higher priority than the others. In the alternatives within these two higher priorities, security and device are the most significant. Since the consistency of the top level is less than 1.0, this set of priorities is considered acceptable.

Table 1: Priority and Consistency Ratio

Criterion	Priority	Alternative	Priority	CR	
				Alt.	Cri.
System Quality	0.028	Online Response Time	0.116	0.05	
		24-hour Availability	0.666		
		Page Loading Speed	0.069		
		Visual Appearance	0.149		
Content Quality	0.124	Up-to-dateness	0.094	0.07	
		Understandability	0.530		
		Timeliness	0.082		
		Preciseness	0.294		
Use	0.046	Information	0.250	0	0.1
		Transaction	0.750		
Trust	*0.430*	*Security*	*0.800*	0	
		Privacy	0.200		
Support	0.060	Tracking Order Status	0.548	0.09	
		Account Maintenance	0.052		
		Payment Alternatives	0.145		
		FAQs	0.256		
Mobility	*0.311*	*Device*	*0.900*	0	
		Application	0.100		

5. Conclusion

We have proposed criteria, for use by a company intending to launch an m-commerce business, by which customer's interests can be assessed. The main attractive factors for the customer are the trust and security factors. In addition, mobility, device, content quality, and the understandability factors are important. Developing aspects of customer interest in the m-commerce market are the main pointers to business success and application of the AHP technique provides a success strategy. Customer views on these criteria are determined by having them complete suitably prepared questionnaires. The AHP technique has application in providing a structured approach to finding the best decision-making strategy in the area of MCDM. If application of these criteria is extended to marketing in the new ubiquitous computing environment, it will become a useful assessment model for predicting and evaluating market tendencies.

References

[1] Efraim Turban, David Kim, Jae Lee, Merrill Warkentin, H. Michael Chung: Electronic Commerce 2002 - A Managerial Perspective. 2nd edn. Prentice Hall, New Jersey(2002).

[2] Ronald R. Yager: Modeling Prioritized Multicriteria Decision Making. IEEE TRANSACTIONS ON SYSTEMS, MAN, AND CYBERNETICS PART B, 34(6), 2396-2404(2004).

[3] Mihai Barbuceanu, Wai-Kau Lo: A Multi-Attribute Utility Theoretic Negotiation Architecture for Electronic Commerce. Proceedings of the fourth international conference on Autonomous agents, Barcelona Spain, 239-246, 2000.

[4] T. L. Saaty: Fundamentals of Decision Making and Priority Theory. 2nd edn. PA: RWS Publications, Pittsburgh(2000).

[5] Alemayehu Molla, Paul S. Licker: E-COMMERCE SYSTEMS SUCCESS: AN ATTEMPT TO EXTEND AND RESPECIFY THE DELONE AND MACLEAN MODEL OF IS SUCCESS. Journal of Electronic Commerce Research, Vol. 2. No. 4. (2001) 131-141.

[6] Peter Tarasewich: Designing Mobile Commerce Applications. COMMUNICATIONS OF THE ACM, Vol. 46. No. 12. (2003) 57-60.

[7] Viswanath Venkatesh, V. Ramesh,Anne P. Massey: Understanding Usability in Mobile Commerce. COMMUNICATIONS OF THE ACM, Vol. 46. No. 12. (2003) 53-56.

[8] Suprateek Sarker, John D. Wells: Understanding Mobile Handheld Device Use and Adap-tion, COMMUNICATIONS OF THE ACM, Vol. 46. No. 12. (2003) 35-40.

[9] Turban, E. & Gehrke, D.: Determinants of e-commerce website. Human Systems Management, 19(2), (2000) 111-120.

[10] Han, K.S. & Noh, M.H.: Critical failure factors that discourage the growth of electronic commerce. International Journal of Electronic Commerce, 4(2) (1999) 25-43.

[11] Von Dran, G.M., Zhang, P., & Small, R.: Quality Websites, An application of the Kano model of Website Desing. In: Haseman, W. D. & Nazareth, D.L. (eds.): Proceedings of the Fifth AMCIS. Association for Information Systems, (1999) 898-901.

[12] Young, D. & Benamati, J.: Differences in public web sites: The current state of large U.S. Firms. Journal of Electronic Commerce Research, 1(3), 94-105(2000).

[13] Warrington, T.B., Abgrab, N.J. & Caldwell, H.M.: Building trust to develop competitive advantage in ebusiness relationships. Competitiveness Review, 10(2) (2000) 160-168.

[14] Kardaras, D. & Karakostas, V.: Measuring the Electronic Commerce Impact on Customer Satisfaction: Experiences, Problems and expectations of the banking sector in the UK. In: Proceeding of the International conference of the Measurement of Electronic Commerce, Singapore (1999).

[15] Adolfo Lozano-Tello, Asuncion Gomez-Perez: BAREMO: how to choose the appropriate software component using the analytic hierarchy process, Proceedings of the 14th international conference on Software engineering and knowledge engineering, Ischia, Italy, 781-788, 2002.

[16] Prakash S Lokachari, Gunavardhan Raj, David D'lima: Selecting the Best e-Commerce Product for a Retail Chain - The Analytic Hierarchy Process Model. Management of Engineering and Technology, Vol. 1. PICMET '01 Portland International Conference, (2001) 145

[17] Zahedi E: The Analytic Hierarchy Process-A Survey of the Method and its Applications. INTERFACES. 16(4) (1986) 96-108.

Brill Academic Publishers
P.O. Box 9000, 2300 PA Leiden,
The Netherlands

Lecture Series on Computer and Computational Sciences
Volume 2, 2005, pp. 88-91

A Posteriori Error Estimation of Quantities of Interest for the Linear Elasticity Problems

S. Korotov[1]

Institute of Mathematics
Helsinki University of Technology
P.O. Box 1100, FI-02015 TKK, Finland

Received 7 April, 2005; accepted in revised form 16 April, 2005

Abstract: The *modus operandi* (technology) presented in the paper is designed to evaluate the accuracy of computed solutions, measured in terms of "problem-oriented" criteria that can be chosen by users. The technology is applicable to various problems in mechanics and physics embracing mathematical models of diffusion processes and models arising in the elasticity theory of solid bodies. Using the linear elasticity model in the present work, we demonstrate that the proposed technology can be easily coded and attached as an independent *programme-checker* to the most of existing educational and industrial codes (such as, e.g., Matlab, ADINA, ANSYS, etc) that use the finite element method as a computational tool.

Keywords: a posteriori error estimation, quantity of interest, finite element method, linear elasticity, superconvergence.

Mathematics Subject Classification: 65N15, 65N30, 65N50

1 Linear Elasticity Model

We consider a classical problem of the linear elasticity theory: Find the displacement field $\mathbf{u} = \mathbf{u}(x_1, ..., x_d) = [u_1, ..., u_d]^T$ of d variables $x_1, ..., x_d$ $(d = 1, 2, ...)$ in the elastic body $\Omega \subset \mathbf{R}^d$ subject to volume and surface forces, described by the following system of equations with boundary conditions $(i = 1, ..., d)$:

$$-\sum_{j,k,l=1}^{d} \frac{\partial}{\partial x_j} (L_{ijkl}\, \varepsilon_{kl}(\mathbf{u})) = f_i \quad \text{in} \quad \Omega, \tag{1}$$

$$u_i = u_0^i \quad \text{on} \quad \Gamma_1, \tag{2a}$$

$$\sum_{j,k,l=1}^{d} n_j\, L_{ijkl}\, \varepsilon_{kl}(\mathbf{u}) = g_i \quad \text{on} \quad \Gamma_2, \tag{2b}$$

where L_{ijkl} are the elastic coefficients, $\varepsilon_{kl}(\mathbf{u})$ are the entries of the strain tensor, and $\Gamma_1 \cup \Gamma_2 = \partial\Omega$.

[1]Corresponding author, e-mail: sergey.korotov@hut.fi

2 Problem-Oriented Criterion

Let $\mathbf{u_h} = \mathbf{u_h}(x_1, ..., x_d) = [u_h^1, ..., u_h^d]^T \in \mathbf{V_h} + \mathbf{u_0}$, where $\mathbf{u_0} = (u_0^1, ..., u_0^d)$, be a standard continuous piecewise affine finite element approximation for the solution of the problem (1)–(2) obtained with help of finite-dimensional space $\mathbf{V_h}$ constructed by means of a selected set of finite element trial (d-dimensional) vector-functions defined on finite element mesh $\mathcal{T}_h$ over Ω. We notice that space $\mathbf{V_h}$ is chosen so that its functions $\mathbf{w_h}$ vanish on Γ_1.

Many users of the existing industrial software are often interested not only in the overall error $\mathbf{e} = \mathbf{u} - \mathbf{u_h}$, but also in its local behaviour, e.g., in a certain subdomain (*zone of interest*) $\omega \subset \Omega$. One way to get an information about such a behaviour is to measure the error in terms of specially selected *problem-oriented criteria*. For example, we can be interested in the following value

$$\int_\Omega \mathbf{\Phi} \cdot (\mathbf{u} - \mathbf{u_h})\, dx, \tag{3}$$

where $\mathbf{\Phi}$ is a selected vector-function such that supp $\mathbf{\Phi} \subseteq \omega$. This type of measuring the error is called in mathematical literature as *measuring error in terms of linear functionals*. Often, is also called *measuring error in quantities of interest*, since the value $\int_\Omega \mathbf{\Phi} \cdot \mathbf{u}\, dx$ can be of a special interest (and that is why it is called the quantity of interest) and we could be interested how far the (computable) value $\int_\Omega \mathbf{\Phi} \cdot \mathbf{u_h}\, dx$ is from this (unknown) quantity of interest.

3 Technology for Error Estimation of Problem-Oriented Criterion (3)

First, two technical steps must be performed. They consist of finding an approximate solution of an auxiliary (the so-called *adjoint*) problem and making a certain post-processing of the solution of this problem, and also of the finite element approximation $\mathbf{u_h}$.

3.1 Auxiliary problem and its finite element solution

Let $\mathbf{V}_\tau$ be another finite-dimensional space constructed by means of a selected set of finite element trial functions on another standard finite element mesh $\mathcal{T}_\tau$ over Ω. We notice that space $\mathbf{V}_\tau$ is chosen so that its functions $\mathbf{w}_\tau$ vanish on Γ_1, and also that $\mathcal{T}_\tau$ need not to coincide with $\mathcal{T}_h$.

The adjoint finite-dimensional problem reads as follows: Find a function $\mathbf{v}_\tau = \mathbf{v}_\tau(x_1, ..., x_d) \in \mathbf{V}_\tau$ such that

$$\int_\Omega \sum_{i,j,k,l=1}^{d} L_{ijkl}\varepsilon_{ij}(\mathbf{v}_\tau)\varepsilon_{kl}(\mathbf{w}_\tau)\, dx = \int_\Omega \mathbf{\Phi} \cdot \mathbf{w}_\tau\, dx \quad \forall \mathbf{w}_\tau \in \mathbf{V}_\tau. \tag{4}$$

3.2 Strain averaging procedures

On $\mathcal{T}_h$, we define the *tensor averaging transformation* $\mathbf{G_h}$ mapping the *tensor of small strains* computed by the finite element approximation $\mathbf{u_h}$

$$\varepsilon(\mathbf{u_h}) = \left[\frac{1}{2}\left(\frac{\partial u_h^i}{\partial x_j} + \frac{\partial u_h^j}{\partial x_i}\right)\right]_{i,j=1}^{d}, \tag{5}$$

which is a constant $d \times d$ matrix over each element of the finite element mesh, into a $d \times d$ matrix-valued continuous piecewise affine function

$$\mathbf{G_h}(\varepsilon(\mathbf{u_h})) = \left[G_h^{i,j}(\varepsilon(\mathbf{u_h}))\right]_{i,j=1}^{d}, \tag{6}$$

(see, e.g. [1] for the exact definition). Similarly, on $\mathcal{T}_\tau$, we define the tensor averaging transformation $\mathbf{G}_\tau$ mapping the tensor of small strains $\varepsilon(\mathbf{v}_\tau)$ into a $d \times d$ matrix-valued continuous piecewise affine function $\mathbf{G}_\tau(\varepsilon(\mathbf{v}_\tau)) = \left[G_\tau^{i,j}(\varepsilon(\mathbf{v}_\tau))\right]_{i,j=1}^{d}$.

3.3 The estimator

We propose to estimate the quantity (3) by the quantity $\mathbf{E}(\mathbf{u_h}, \mathbf{v}_\tau)$ given by the following formula:

$$\mathbf{E}(\mathbf{u_h}, \mathbf{v}_\tau) := \mathbf{E_0}(\mathbf{u_h}, \mathbf{v}_\tau) + \mathbf{E_1}(\mathbf{u_h}, \mathbf{v}_\tau), \tag{7}$$

where

$$\mathbf{E_0}(\mathbf{u_h}, \mathbf{v}_\tau) = \int_\Omega \mathbf{f} \cdot \mathbf{v}_\tau \, dx + \int_{\Gamma_2} \mathbf{g} \cdot \mathbf{v}_\tau \, ds - \int_\Omega \sum_{i,j,k,l=1}^{d} L_{ijkl} \varepsilon_{ij}(\mathbf{u_h}) \varepsilon_{kl}(\mathbf{v}_\tau) \, dx, \tag{8}$$

and

$$\mathbf{E_1}(\mathbf{u_h}, \mathbf{v}_\tau) = \int_\Omega \sum_{i,j,k,l=1}^{d} L_{ijkl} \left(\varepsilon_{ij}(\mathbf{u_h}) - G_h^{i,j}(\varepsilon(\mathbf{u_h}))\right) \left(\varepsilon_{kl}(\mathbf{v}_\tau) - G_\tau^{k,l}(\varepsilon(\mathbf{v}_\tau))\right) dx. \tag{9}$$

The quantity $\mathbf{E}(\mathbf{u_h}, \mathbf{v}_\tau)$ is directly computable once the approximations $\mathbf{u_h}$ and $\mathbf{v}_\tau$ are found. The procedure of construction of the estimator is similar to the construction of the estimator for the case of linear elliptic problem of the second order described in [2] and is based on the phenomenon of the superconvergence of the averaged gradients, see [1].

4 Numerical Experiments

We deal with the plane stress problem with the following parameters: $E = 10^6$, $\nu = 0.3$, and $\mathbf{f} = (0, 0)$. The PDE Toolbox of Matlab is used for the mesh generation purposes. The solution domain Ω and the zone of interest ω are presented in Fig. 1. The Dirihlet conditions $\mathbf{u_0} = (0, \pm 1)$ are prescribed on the upper and the lower parts of the domain Ω, respectively. In the remaining parts of $\partial\Omega$ we impose the homogeneous Neumann conditions. The finite element solution $\mathbf{u_h}$ calculated on the mesh $\mathcal{T}_h$ with 148 nodes and the corresponding von Mises stress distribution are given in Fig. 1.

The function $\mathbf{\Phi}$ is prescribed to be (1, 1) in ω and (0, 0) outside of it. The results of calculations for the estimator are presented in Table 1. The first and the second columns contain the numbers of nodes in the primal ($\mathcal{T}_h$) and adjoint ($\mathcal{T}_\tau$) meshes, respectively. Two typical adjoint meshes are also given in Fig. 1. The symbol I_{eff} denotes the so-called *effectivity index*

$$I_{eff} = \frac{|\mathbf{E}(\mathbf{u_h}, \mathbf{v}_\tau)|}{|\int_\Omega \mathbf{\Phi} \cdot (\mathbf{u} - \mathbf{u_h}) \, dx|}. \tag{10}$$

The tests performed clearly demonstrate the fact that even in the presence of the strong singularity in the solution (at the point (0, 0)), an usage of a computationally cheap adjoint mesh with a considerable smaller amount of nodes with respect to the number of nodes in the primal problem is effective. The performance of the estimator is, of course, becoming considerably better if we solve problems with weaker singularities in solutions, and also in the situations when we have enough computer facilities (time, memory, etc) and can solve adjoint problems on very dense adjoint meshes.

Table 1: The results of performance of the estimator $\mathbf{E}$

Prim	Adj	$\mathbf{E_0}$	$\mathbf{E_1}$	$\mathbf{E}$	$\int_\Omega \mathbf{\Phi} \cdot (\mathbf{u} - \mathbf{u_h})\, dx$	I_{eff}
148	75	0.000594	0.000417	0.001012	0.001516	0.67
148	77	0.000565	0.000558	0.001123	0.001516	0.74
148	90	0.000696	0.000510	0.001206	0.001516	0.80
148	123	0.000852	0.000425	0.001277	0.001516	0.84
148	226	0.000957	0.000355	0.001312	0.001516	0.87

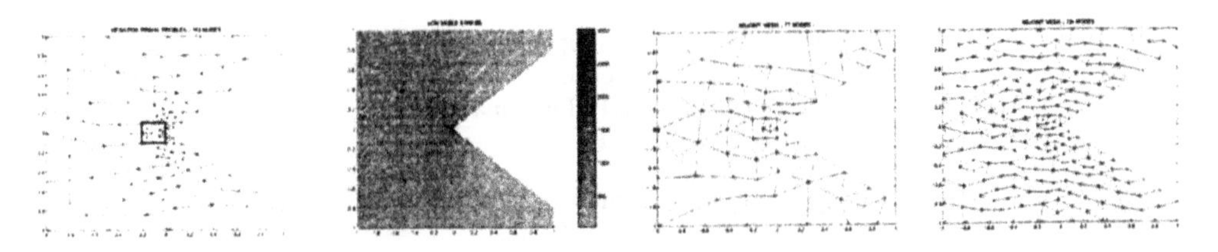

Figure 1: Solution domain Ω, zone of interest ω, mesh $\mathcal{T}_h$ (first), von Mises stress (second), adjoint mesh $\mathcal{T}_\tau$ with 77 nodes (third) and with 226 nodes (fourth).

5 Comments

The effectivity of the proposed technique, strongly increases when one is interested not in a single solution of the primal problem for a concrete data, but analyzes a series of approximate solutions for a certain set of boundary conditions and various right-hand sides (which is typical in the engineering design when it is necessary to model the behavior of a construction for various working regimes). In this case, the adjoint problem must be solved *only once* for each "quantity". In particular, we can compute $\mathbf{v}_\tau$ on very dense (adjoint) mesh, and use it further in testing the accuracy of approximate solutions of various relevant primal problems.

Acknowledgment

The author was supported by the Academy Research Fellowship no. 208628 from the Academy of Finland.

References

[1] I. Hlaváček, M. Křížek. On a superconvergent finite element scheme for elliptic systems. I. Dirichlet boundary conditions. *Apl. Mat.* 32, 131 154, 1987.

[2] S. Korotov, P. Neittaanmäki, S. Repin. A posteriori error estimation of goal-oriented quantities by the superconvergence patch recovery. *J. Numer. Math.* 11, 33 59, 2003.

Brill Academic Publishers
P.O. Box 9000, 2300 PA Leiden,
The Netherlands

Lecture Series on Computer and Computational Sciences
Volume 2, 2005, pp. 92-96

If I know it then it can't be false (and if it's true then it is not impossible)

Costas D. Koutras[1] **Christos Nomikos**[2] **Pavlos Peppas**[3]

[1]Department of Computer Science and Technology
University of Peloponnese
End of Karaiskaki Str., 22100 Tripolis, Greece
ckoutras@uop.gr

[2]Department of Computer Science
University of Ioannina
P.O. Box 1186, 45 110 Ioannina, Greece
cnomikos@cs.uoi.gr

[3]Department of Business Administration
University of Patras
Patras 265 00, Greece
peppas@otenet.gr

Received April 10, 2005; accepted in revised form April 26, 2005

Abstract: In this paper we describe a three-valued logic of belief, a member of the family of many-valued modal logics introduced by M. Fitting. This logic possesses the axioms of positive and negative introspection along with an interesting version of axiom **T** which asserts that a known fact cannot be false, even if it has not been verified yet. The proposed logic is of computational interest for two reasons (i) it can also be seen as the logic describing the epistemic consensus of two interrelated agents: a **K45** agent dominating an **S5** agent, and (ii) its satisfiability problem is NP-complete. The latter should be contrasted to the well-known fact that the satisfiability problem for the above mentioned epistemic logics with two or more agents is PSPACE-complete.

Keywords: Modal logic, epistemic logic, multi-agent systems.

Mathematics Subject Classification: 68T27 (Logic in artificial intelligence), 68T30 (Knowledge representation), 03B42 (Logic of knowledge and belief)

1 Introduction and Motivation

Modal epistemic logics have been introduced in the field of Philosophical Logic but have found very interesting applications in Computer Science and Artificial Intelligence [1, 5]. Some of the most important axioms in epistemic logic are: the ubiquitus **K**. $\Box(X \supset Y) \wedge \Box X \supset \Box Y$ (indicating that the logical consequences of knowledge constitute knowledge), **D**. $\Box X \supset \Diamond X$ (consistent belief: it is not the case that contradictory facts are known), **4**. $\Box X \supset \Box\Box X$ (positive introspection: what is known, is known to be known), **5**. $\neg\Box X \supset \Box\neg\Box X$ (negative introspection) and **T**. $\Box X \supset X$ (knowledge as true, justified belief: what is known is true)[1]. The last axiom is usually considered as the one that differentiates 'knowledge' from 'belief' and has been widely criticized. Some of the most important axiomatic systems in epistemic logic, have been **K45**, its 'consistent-belief' version **KD45** and **KT45** (a strong logic that describes an unrealistically ideal introspective reasoner), which carries the traditional name **S5**. For the readers not already acquainted with modal logic, it should be obvious that the axiom systems are named after the (names of) axioms they adopt.

The explosion of interest into multi-agent systems has placed new interesting problems in Knowledge Representation, as the new computing environments can be modelled as teams of intelligent entities acting and cooperating in complex environments. A decade ago, M. Fitting introduced in a series of paper, a family of many-valued modal logics which possess an interesting equivalent interpretation with respect to multi-agent applications [3, 4]. This family of logics has been further explored in [6, 7, 2] from the model-theoretic perspective. In this paper we describe a concrete example of this family of logics, its axiomatic content, and present completeness and complexity results. We start with a motivating example.

Assume the following scenario: We have two intelligent agents in our application domain, whose opinion we value. Both agents $\mathcal{A}$ and $\mathcal{B}$ have a knowledge of the environment in which they act, encoded in a common epistemic language; we may think of a modal formula $\Box X$ as saying that "X is known to be true, under all possible circumstances". Moreover there exist (possibly infinitely) many situations that might be of interest and it is natural (and quite common) to think of them as "possible worlds". We wish to be able to calculate efficiently the "consensus" of the two agents, that is, the epistemic formulae in which the two agents agree in every possible situation, even though each agent may have a different view on how do these situations relate to each other. It is not difficult to check that if the agents $\mathcal{A}$ and $\mathcal{B}$ are completely independent, there is nothing genuinely novel and interesting in this scenario (see [4]). What might make the situation interesting is the additional constraint that agent $\mathcal{A}$ "dominates" agent $\mathcal{B}$: if $\mathcal{A}$ assumes X is true (X is an atomic formula) then $\mathcal{B}$ will necessarily assume so. In this case let us impose the following rules in the recursive definition of truth assignment for the formulae of the common logical language: conjunctions and disjunctions are evaluated "locally" (their truth value does not depend on others' opinion), but for agent $\mathcal{A}$ to consider an implication $X \supset Y$ or a negation $\neg X$ to be true, also the "dominated" agent $\mathcal{B}$ should say so. What about the modal formulae $\Box X$ and $\Diamond X$? We assume that each agent has a different Kripke model in mind: the two models share the same set of possible worlds and their accessibility relations comply with the "domination" requirements sketched above: if $\mathcal{A}$ (the "dominating" agent) considers world t an alternative to world s, agent $\mathcal{B}$ (the "dominated" one) says so. In that way, the evaluation of $\Box X$ and $\Diamond X$ reflects the dominance relation between the agents: a pictorial example follows.

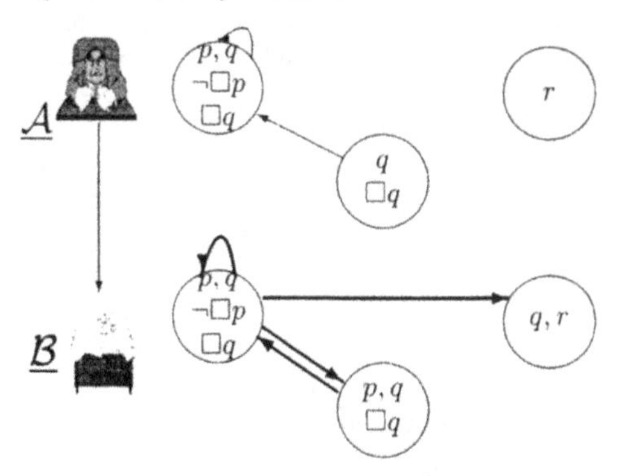

What can we say about the formulae on which our agents agree? The problem can be reformulated into a many valued epistemic logic, assuming "admissible" sets of experts as generalized truth values. It suffices to notice that only $\emptyset$, $\{\mathcal{B}\}$ and $\{\mathcal{A}, \mathcal{B}\}$ can serve as truth values, as it impossible to have a situation in which only $\mathcal{A}$ assumes X to be true. Thus, the above picture transforms as follows, under the interpretation mentioned. Thus, this agent-based scenario provides us with an interesting three-valued modal logic, which may become even more interesting if we assume that $\mathcal{A}$ is an agent reasoning in terms of the modal doxastic logic **K45** while $\mathcal{B}$ is an idealized epistemic reasoner with full introspection capabilities (an **S5** agent).

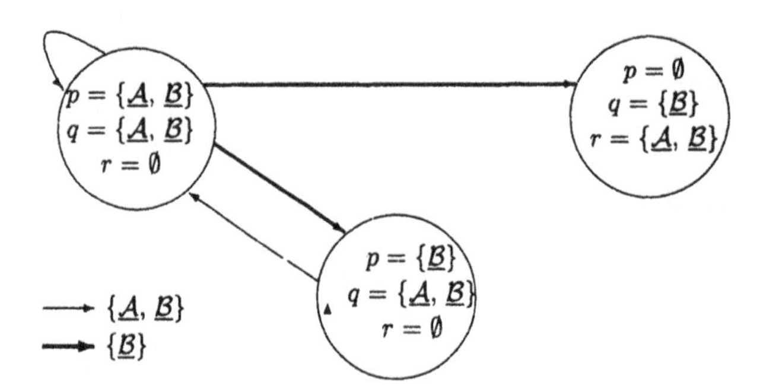

In that case $\mathcal{A}$ has in mind a serial, transitive, euclidean frame, while $\mathcal{B}$ has in mind a frame whose accessibility relation is an equivalence relation ([1]). In that case we come up with an interesting three-valued epistemic logic, equipped with the introspection axioms **4** and **5**, along with a variant of axiom $\mathbf{T}^h$. $\Box X \supset (\mathsf{h} \supset X)$ naturally interpreted as *"if X is known then it can't be false"* and a dual axiom $\mathbf{T}^h_d$. $X \supset (\mathsf{h} \supset \Diamond X)$ interpreted as *"If X is true then it can't be impossible"*. One of the strongest motivations for studying this logic is that, as proved below, its satisfiability problem is NP-complete, not harder than the relevant problem of **KD45** and **S5**, its "constituent" logics. This implies that computing the consensus of the agents might be cheaper than computing multi agent knowledge, as it is known that the satisfiability problem for epistemic logic with at least two agents $\mathbf{KD45}_n$, $\mathbf{S5}_n$ $(n \geq 2)$ is PSPACE-complete [5].

2 Syntax, Semantics, Proof Theory

To define the modal language of our three-valued modal logic we first fix the space of truth values: $\mathcal{H}_3 = \{\mathsf{f}, \mathsf{h}, \mathsf{t}\}$, ordered linearly under $\mathsf{f} \leq \mathsf{h} \leq \mathsf{t}$ (f and t stand for falsity and truth respectively, while h stands for an intermediate truth value (half)). It is known that every finite linear lattice is a Heyting algebra equipped with the operations $a \wedge b = \min(a, b)$, $a \vee b = \max(a, b)$ and the relative pseudo-complement $a \Rightarrow b$ which is t, if $a \leq b$ and it equals b otherwise (see [9] for all details about Heyting algebras and their properties). The elements of $\mathcal{H}_3$ are directly represented in the language and Φ denotes a set of of *propositional variables*. The **syntax** of the modal language $L^3_{\Box\Diamond}(\Phi)$ is given by (a ranges over elements of $\mathcal{H}_3$, P ranges over elements of Φ):

$$A ::= a \mid P \mid A_1 \vee A_2 \mid A_1 \wedge A_2 \mid A_1 \supset A_2 \mid \Box A \mid \Diamond A$$

$(A_1 \equiv A_2)$ abbreviates $(A_1 \supset A_2) \wedge (A_2 \supset A_1)$ and $\neg A$ abbreviates $(A \supset \mathsf{f})$. The **semantics** of our three-valued modal languages are provided by the definition below, which introduces possible-world models with a three-valued accessibility relation.

Definition 1 ([4]) An $\mathcal{H}_3$-*modal frame* for $L^3_{\Box\Diamond}(\Phi)$ is a pair $\mathfrak{F} = \langle \mathfrak{S}, \mathfrak{g} \rangle$, where $\mathfrak{S}$ is a non-empty set of *states* and $\mathfrak{g} : \mathfrak{S} \times \mathfrak{S} \to \mathcal{H}_3$ is a total function mapping pairs of states to elements of $\mathcal{H}_3$.
An $\mathcal{H}_3$-*modal model* $\mathfrak{M} = \langle \mathfrak{S}, \mathfrak{g}, v \rangle$ is built on $\mathfrak{F}$ by providing a *valuation* v, that is a function

$v : \mathfrak{S} \times (\mathcal{H} \cup \Phi) \to \mathcal{H}_3$ which assigns an $\mathcal{H}_3$-*truth value* to atomic formulas in each state, such that $v(\mathfrak{s}, a) = a$, for every $\mathfrak{s} \in \mathfrak{S}$ and $a \in \mathcal{H}_3$. That is, propositional constants are always mapped to 'themselves'.

The following definition extends the valuation v to all the formulas of $L^3_{\Box\Diamond}(\Phi)$ in a standard recursive fashion.

Definition 2 Let $\mathfrak{M} = \langle \mathfrak{S}, \mathfrak{g}, v \rangle$ be an $\mathcal{H}_3$-*model*. The extension of the valuation v is given by the following clauses: $v(\mathfrak{s}, A \wedge B) = \min\Big(v(\mathfrak{s}, A), v(\mathfrak{s}, B)\Big)$, $v(\mathfrak{s}, A \vee B) = \max\Big(v(\mathfrak{s}, A), v(\mathfrak{s}, B)\Big)$ $v(\mathfrak{s}, A \supset B) = \Big(v(\mathfrak{s}, A) \Rightarrow v(\mathfrak{s}, B)\Big)$,$v(\mathfrak{s}, A \supset B) = \Big(v(\mathfrak{s}, A) \Rightarrow v(\mathfrak{s}, B)\Big)$ for the propositional connectives and for the modal operators:

- $v(\mathfrak{s}, \Box A) \;=\; \min_{\mathfrak{t} \in \mathfrak{S}} \Big(\mathfrak{g}(\mathfrak{s}, \mathfrak{t}) \Rightarrow v(\mathfrak{t}, A)\Big)$
- $v(\mathfrak{s}, \Diamond A) \;=\; \max_{\mathfrak{t} \in \mathfrak{S}} \Big(\min\Big(\mathfrak{g}(\mathfrak{s}, \mathfrak{t}), v(\mathfrak{t}, A)\Big)\Big)$

The semantics described below alters in a very interesting manner the notion of relational frames, in that it concerns a kind of three-valued accessibility relation. The frame can be seen as (possibly infinite) directed graph, in which every edge is either grey or black. If the accessibility relation has value h (t) for a pair of states, then these states are connected by a grey (resp. black edge) in the graph. If the accessibility relation has value f for a pair of states, then these states are not connected; it can be checked by the definition above that edges labelled f do not play a role in the calculation of truth values. The **proof theory** of this modal logic, comprises an interesting Gentzen-style sequent calculus (see [4, 6]), which is omitted here due to space limitations. We can now described the promised three valued modal epistemic logic: its axiomatization comprises $\mathbf{K} : \Box(A \supset B) \supset (\Box A \supset \Box B)$, $\mathbf{4} : \Box X \supset \Box\Box X$, $\mathbf{4}_d : \Diamond\Diamond X \supset \Diamond X$, $\mathbf{5} : \Diamond X \supset \Box\Diamond X$, $\mathbf{5}_d : \Diamond\Box X \supset \Box X$, and the new axioms of "not-fully-justified" belief

$$\mathbf{T}^h : \Box X \supset (\mathsf{h} \supset X) \quad \text{and} \quad \mathbf{T}^h_d : X \supset (\mathsf{h} \supset \Diamond X)$$

The following **completeness theorem** follows essentially from the frame determination results of [6]; it can be given a simpler presentation for the case of three, linearly-ordered, truth values but the details are left for the full paper. It is however this result that allows us to derive the complexity characterization of the next section.

Theorem 2.1 *The logic axiomatized from the axioms above, is sound and complete with respect to all three-valued frames that satisfy a natural many-valued analog of transitivity* $(\mathfrak{g}(\mathfrak{s}, \mathfrak{t}) \wedge \mathfrak{g}(\mathfrak{t}, \mathfrak{r}) \leq \mathfrak{g}(\mathfrak{s}, \mathfrak{r}))$ *and euclideanness* $(\mathfrak{g}(\mathfrak{s}, \mathfrak{t}) \wedge \mathfrak{g}(\mathfrak{s}, \mathfrak{r}) \leq \mathfrak{g}(\mathfrak{t}, \mathfrak{r}))$, *while each state has a grey (*h*-labelled) self-loop.*

The above mentioned algebraic properies of modal frames can be easily described in terms of edge coloring (black-grey); perhaps the reader wishes to try. These properties allow us to construct a "small" model for each $L^3_{\Box\Diamond}(\Phi)$ that is consistent with our logic, that is, it can be satisfied in a state of a model based on an appropriate three-valued frame.

3 Complexity

In this section we prove that our logic has an NP-complete satisfiability problem, which implies that *computing the epistemic agreement might be cheaper than computing knowledge*!

Theorem 3.1 *Given a formula A of* $L^3_{\Box\Diamond}$, *it is NP-complete to decide if it is satisfiable in an* $\mathcal{H}_3$-*modal model of the kind described in Theorem 2.1.*

PROOF. (*Sketch*) Due to space limitations the details of the proof are left for the full paper. The NP-hardess result, is obtained by a simple reduction from the SAT problem: it is easy to check that a formula ϕ is satisfiable in the propositional two-valued sense, if an only if it is satisfiable in an arbitrary $\mathcal{H}_3$-modal model.

In order to prove that the satisfiability problem for our three valued logic is in NP we prove two facts: (i) if A is satisfied in a $\mathcal{H}_3$-modal model based on a frame of the type we are interested in, then it is satified in a $\mathcal{H}_3$-modal model with the same properties, with at most $|A|$ states, that is with size polynomial to the size of A, and (ii) given an $\mathcal{H}_3$-modal model and a formula in $L^3_{\Box\Diamond}$, the the truth value of the formula for each state can be computed in polynomial time. Thus, a non-deterministic polynomial time algorithm for the satisfiability problem in our three-valued logic starts by non-deterministically costructing a $\mathcal{H}_3$-modal model for A, and then it computes the truth values of A for every state of the model. If the truth value of A is t at some state, then it returns 'yes', otherwise it returns 'no'. The details of the proof are a bit lengthy and will be given in the full paper. ∎

4 Conclusions

We have presented a three-valued logic, which has been actually inspired from a question posed by M. Fitting in the epilogue of [4]. It is interesting to note that the modal (and propositional constant) free part of this logic is actually a widely-used intermediate logic, the so called *"logic of here and there"* which has been employed in the study of Non-Monotonic reasoning and logic programming [8].

References

[1] P. Blackburn, M. de Rijke, and Y. Venema. *Modal Logic*. Number 53 in Cambridge Tracts in Theoretical Computer Science. Cambridge University Press, 2001.

[2] P. Eleftheriou and C. D. Koutras. Frame constructions, truth invariance and validity preservation in many-valued modal logic. Submitted, 2004.

[3] M. C. Fitting. Many-valued modal logics. *Fundamenta Informaticae*, 15:235–254, 1991.

[4] M. C. Fitting. Many-valued modal logics II. *Fundamenta Informaticae*, 17:55–73, 1992.

[5] J. Halpern and Y. Moses. A guide to completeness and complexity for modal logics of knowledge and belief. *Artificial Intelligence*, 54(2):319–379, 1992.

[6] C. D. Koutras, Ch. Nomikos, and P. Peppas. Canonicity and completeness results for many-valued modal logics. *Journal of Applied Non-Classical Logics*, 12(1):7–42, 2002.

[7] C. D. Koutras. A catalog of weak many-valued modal axioms and their corresponding frame classes. *Journal of Applied Non-Classical Logics*, 13(1):47–71, 2003.

[8] D. Pearce. Stable inference as intuitionistic validity. *Journal of Logic Programming*, 38(1):79–91, 1999.

[9] H. Rasiowa and R. Sikorski. *The Mathematics of Metamathematics*. PWN - Polish Scientific Publishers, Warsaw, third edition, 1970.

Brill Academic Publishers
P.O. Box 9000, 2300 PA Leiden,
The Netherlands

Lecture Series on Computer and Computational Sciences
Volume 2, 2005, pp. 97-101

A Multistep Block Scheme for the Numerical Solution of Ordinary Differential Equations

G. Psihoyios[1]

Department of Mathematics and Technology,
Faculty of Applied Sciences,
Anglia Polytechnic University,
Cambridge CB1 1PT, UK

Received 10 April, 2004; accepted in revised form 24 April, 2005

Abstract: In this work we propose a new multistep block scheme, which has been constructed with the aim to successfully tackle the numerical solution of certain types of ordinary differential equations. In order to maximise the chances of producing a good new method, the first mandatory step is to ensure it incorporates the best possible theoretical properties. We have prepared a rather detailed account of the new methods and space restrictions allowing, the properties of the scheme will be discussed in some length.

Keywords: Block methods, Multistep methods, Stability, Initial value problems.

Mathematics Subject Classification: 65L05, 65L06, 65L07, 65L12, 65L20

1. Introduction

It is more than thirty years that various types of block methods have appeared for the numerical solution of stiff and non-stiff systems of ordinary differential equations (ODEs). Furthermore, the colleagues who created the numerous block-type algorithms have, in many instances, approached the matter in quite dissimilar ways. For example some of these schemes are applied in a cyclic order, some schemes are explicit while others are implicit, some of these algorithms are based on Runge-Kutta methods, other algorithms are based on multistep formulae and some belong to a "hybrid" class of methods, and so on. A very concise and indicative list of related references may be found in [1] through [8] (and references therein).

The new block methods presented here are based on multistep formulae and have been designed for the numerical solution of stiff ODEs. The construction of a block scheme with good theoretical properties is by no means straightforward and in the following paragraphs we will briefly describe some of the problems encountered in the development of the new methods.

2. In Search for a Suitable Block Scheme

The basic idea is to identify two suitable predictors, let us symbolize them as $\bar{y}_{n+k}$ and $\bar{y}_{n+k+1}$, of order k, then to obtain a corrected solution $\hat{y}_{n+k}$ of order k+1 and finally to close the block with a second corrected solution $\hat{y}_{n+k+1}$ also of order k+1. The next block, using past information obtained from the corrected solutions, will start with predictors $\bar{y}_{n+k+2}$ and $\bar{y}_{n+k+3}$ and then the corrected solutions symbolized by $\hat{y}_{n+k+2}$ and $\hat{y}_{n+k+3}$ will be computed. In the same way the subsequent block will be computed, and so on. The new methods are formed essentially from one block. For the sake of convenience we will refer to the first block method (order 2) as *Block* 1, to the subsequent block (order

[1] Correspondence address: 192 Campkin Road, Cambridge CB4 2LH, United Kingdom. E-mail: g.psihoyios@ntlworld.com

3) as *Block* 2 etc. The methods go up to and including order 5, i.e. there are 4 blocks. The reason for adopting this idea is because we anticipate it will permit a relatively cheap, in computational terms, numerical evaluation since the minimum number of function evaluations per two (accepted) steps is four, or a minimum of two function evaluations per step (this depends on the rate of convergence of Newton iterations). Each block advances two integration steps:

A realistic scheme based on the above idea is the following (we will call it *Scheme A*):

- Use a standard backward differentiation formula (BDF) [9] to compute the two predicted solutions of order k.
- Use an Modified Extended BDF (MEBDF) [10] with $\hat{b}_k = b_k - \bar{b}_k$ to compute the first corrected solutions.
- Use a BDF formula (with a slight modification on its right hand side) of order $k+1$ to compute the second corrected solutions.

Such a scheme for block 1 and $k = 1$ takes the form:

$$\textit{1st Predictor}: \quad \bar{y}_{n+1} = y_n + h\bar{y}'_{n+1}$$

$$\textit{2nd Predictor}: \quad \bar{y}_{n+2} = \bar{y}_{n+1} + h\bar{y}'_{n+2}$$

$$\textit{1st Corrector}: \quad \hat{y}_{n+1} - y_n = h\left(-\frac{1}{2}\bar{y}'_{n+2} + \hat{y}'_{n+1} + \frac{1}{2}\bar{y}'_{n+1}\right).$$

$$\textit{2nd Corrector}: \quad \hat{y}_{n+2} - \frac{4}{3}\hat{y}_{n+1} + \frac{1}{3}y_n = h\left(\hat{y}'_{n+2} - \frac{1}{3}\bar{y}'_{n+2}\right)$$

Note that a slight change (which does not affect the left-hand-side known BDF coefficients) of the standard BDF has occurred in the right-hand-side of the second corrector. The right-hand-side should normally be $h\frac{2}{3}\hat{y}'_{n+2}$, but has instead been replaced by $h\left(1 \cdot \hat{y}'_{n+2} + \left(\frac{2}{3} - 1\right)\bar{y}'_{n+2}\right)$.

The above block 1, $k = 1$, scheme causes no problems as far as stability is concerned and both correctors are A-stable [9]. However, when we consider the corresponding block 2, $k = 2$, scheme, the second corrector is NOT necessarily zero-stable [9] any longer. The $k = 2$ scheme is as follows:

$$\textit{1st Predictor}: \quad \bar{y}_{n+3} = \frac{4}{3}y_{n+2} - \frac{1}{3}y_{n+1} + \frac{2}{3}h\bar{y}'_{n+3}$$

$$\textit{2nd Predictor}: \quad \bar{y}_{n+4} = \frac{4}{3}\bar{y}_{n+3} - \frac{1}{3}y_{n+2} + \frac{2}{3}h\bar{y}'_{n+4}$$

$$\textit{1st Corrector}: \hat{y}_{n+3} - \frac{28}{23}y_{n+2} + \frac{5}{23}y_{n+1} = h\left(-\frac{4}{23}\bar{y}'_{n+4} + \frac{2}{3}\hat{y}'_{n+3} + \frac{20}{69}\bar{y}'_{n+3}\right) \quad (1)$$

$$\textit{2nd Corrector}: \hat{y}_{n+4} - \frac{18}{11}\hat{y}_{n+3} + \frac{27}{33}y_{n+2} - \frac{6}{33}y_{n+1} = h\left(\frac{2}{3}\hat{y}'_{n+4} - \frac{4}{33}\bar{y}'_{n+4}\right)$$

If we take the first characteristic polynomial of the 2nd corrector we have:

$$r^4 - \frac{18}{11}r^3 + \frac{27}{33}r^2 - \frac{6}{33}r = 0$$

$$\Rightarrow r\left(r^3 - \frac{18}{11}r^2 + \frac{27}{33}r - \frac{6}{33}\right) = 0 \quad (2)$$

Equation (2) is zero-stable, as expected since it results from BDF. If, however, we substitute in (2) the characteristic polynomial from the first corrector (1), as we should, we obtain:

$$r^3 - \frac{28}{23}r^2 + \frac{5}{23}r = 0 \quad \text{or} \quad r^2 = \frac{28}{23}r - \frac{5}{23}$$

then (2) becomes:

$$r^3 - \frac{18}{11}\left(\frac{28}{23}r - \frac{5}{23}\right) + \frac{27}{33}r - \frac{6}{33} = 0.$$

The above characteristic polynomial is *not* zero-stable any more since one of its roots is > 1 in modulus: $r_1 = 1$, $r_2 = -1.151$ and $r_3 = 0.151$. We also checked different forms of *Scheme A* like:

- The two predictors $\bar{y}_{n+3}$ and $\bar{y}_{n+4}$ to be of order 3 and
- The two correctors $\hat{y}_{n+3}$ and $\hat{y}_{n+4}$ to be of order 4.

None of the above modifications produced a zero-stable second corrector. Thus, the only reasonable course of action left to us was to change the initial *Scheme A*.

3. A Block Scheme for Stiff Differential Equations

After considerable work and testing we finally came up with the following scheme for our new block methods:

(1) Compute the two predicted solutions $\bar{y}_{n+k}$ and $\bar{y}_{n+k+1}$ using a standard BDF of order k (same as before).

(2) Compute the first corrected solution $\hat{y}_{n+k}$ using an implicit multistep formula of order k+1.

(3) Close the block by computing the second corrected solution $\hat{y}_{n+k+1}$ using a standard BDF of order k+1 (same as before).

(4) There will be a total of 4 blocks, $\tau = 1,2,3,4$. These blocks are of algebraic orders between 2 and 5 respectively (same as before).

As can be seen, compared to our initial *Scheme A*, we have only changed the formula for computing the first corrector, which proved to be crucial for the entire algorithm. We now present the 4 blocks of the block methods in detail:

Block 1. The predictors are of order 1 and the correctors are of order 2.

Predictor 1, $\bar{y}_{n+k} - y_{n+k-1} = h\bar{y}'_{n+k}$ (3a)

Predictor 2, $\bar{y}_{n+k+1} - \bar{y}_{n+k} = h\bar{y}'_{n+k+1}$ (3b)

$$\textit{Corrector 1}, \ \hat{y}_{n+k} - y_{n+k-1} = h\left[\left(\frac{3}{4} - \frac{\bar{b}_k}{2}\right)\bar{y}'_{n+k+1} + \bar{b}_k\hat{y}'_{n+k} + \left(\frac{1}{4} - \frac{\bar{b}_k}{2}\right)y'_{n+k-1}\right] \quad (3c)$$

$$\textit{Corrector 2}, \ \hat{y}_{n+k+1} - \frac{4}{3}\hat{y}_{n+k} + \frac{1}{3}y_{n+k-1} = h\left[\hat{y}'_{n+k+1} - \frac{1}{3}\bar{y}'_{n+k+1}\right] \quad (3d)$$

Block 2: The predictors are of order 2 and the correctors are of order 3.

$$\textit{Predictor 1}, \ \bar{y}_{n+k} - \frac{4}{3}y_{n+k-1} + \frac{1}{3}y_{n+k-2} = \frac{2}{3}h\bar{y}'_{n+k} \quad (4a)$$

$$\text{Predictor 2.}\quad \bar{y}_{n+k+1} - \frac{4}{3}\bar{y}_{n+k} + \frac{1}{3}y_{n+k-1} = \frac{2}{3}h\bar{y}'_{n+k+1} \tag{4b}$$

$$\text{Corrector 1.}\quad \hat{y}_{n+k} + \left(-\frac{1}{2} - \frac{3}{4}\bar{b}_k\right)y_{n+k-1} + \left(\frac{3}{4}\bar{b}_k - \frac{1}{2}\right)y_{n+k-2}$$
$$= h\left[\left(\frac{1}{8} - \frac{5}{16}\bar{b}_k\right)\bar{y}'_{n+k+1} + \bar{b}_k\hat{y}'_{n+k} + \left(\frac{11}{8} - \frac{23}{16}\bar{b}_k\right)y'_{n+k-1}\right] \tag{4c}$$

$$\text{Corrector 2.}\quad \hat{y}_{n+k+1} - \frac{18}{11}\hat{y}_{n+k} + \frac{27}{33}y_{n+k-1} - \frac{6}{33}y_{n+k-2}$$
$$= h\left[\frac{2}{3}\hat{y}'_{n+k+1} - \frac{4}{33}\bar{y}'_{n+k+1}\right] \tag{4d}$$

<u>Block 3</u> : The predictors are of order 3 and the correctors are of order 4.

$$\text{Predictor 1.}\quad \bar{y}_{n+k} - \frac{18}{11}y_{n+k-1} + \frac{27}{33}y_{n+k-2} - \frac{6}{33}y_{n+k-3} = \frac{6}{11}h\bar{y}'_{n+k} \tag{5a}$$

$$\text{Predictor 2.}\quad \bar{y}_{n+k+1} - \frac{18}{11}\bar{y}_{n+k} + \frac{27}{33}y_{n+k-1} - \frac{6}{33}y_{n+k-2} = \frac{6}{11}h\bar{y}'_{n+k+1} \tag{5b}$$

$$\text{Corrector 1.}\quad \hat{y}_{n+k} + \left(\frac{9}{38} - \frac{165}{76}\bar{b}_k\right)y_{n+k-1} + \left(-\frac{27}{19} + \frac{48}{19}\bar{b}_k\right)y_{n+k-2}$$
$$+\left(\frac{7}{38} - \frac{27}{76}\bar{b}_k\right)y_{n+k-3} = h\left[\left(\frac{3}{38} - \frac{17}{76}\bar{b}_k\right)\bar{y}'_{n+k+1}\right.$$
$$\left. +\bar{b}_k\hat{y}'_{n+k} + \left(\frac{75}{38} - \frac{197}{76}\bar{b}_k\right)y'_{n+k-1}\right] \tag{5c}$$

$$\text{Corrector 2.}\quad \hat{y}_{n+k+1} - \frac{48}{25}\hat{y}_{n+k} + \frac{36}{25}y_{n+k-1} - \frac{16}{25}y_{n+k-2} + \frac{3}{25}y_{n+k-3}$$
$$= h\left[\frac{6}{11}\hat{y}'_{n+k+1} - \frac{18}{275}\bar{y}'_{n+k+1}\right] \tag{5d}$$

<u>Block 4</u> : The predictors are of order 4 and the correctors are of order 5.

$$\text{Predictor 1.}\quad \bar{y}_{n+k} - \frac{48}{25}y_{n+k-1} + \frac{36}{25}y_{n+k-2} - \frac{16}{25}y_{n+k-3} + \frac{3}{25}y_{n+k-4} = \frac{12}{25}h\bar{y}'_{n+k} \tag{6a}$$

$$\text{Predictor 2.}\quad \bar{y}_{n+k+1} - \frac{48}{25}\bar{y}_{n+k} + \frac{36}{25}y_{n+k-1} - \frac{16}{25}y_{n+k-2} + \frac{3}{25}y_{n+k-3} = \frac{12}{25}h\bar{y}'_{n+k+1} \tag{6b}$$

$$\text{Corrector 1.}\quad \hat{y}_{n+k} + \left(\frac{1092 - 4009\bar{b}_k}{963}\right)y_{n+k-1} + \left(\frac{-588 + 1171\bar{b}_k}{214}\right)y_{n+k-2}$$
$$+\left(\frac{76 - 163\bar{b}_k}{107}\right)y_{n+k-3} + \left(\frac{-186 + 413\bar{b}_k}{1926}\right)y_{n+k-4}$$
$$= h\left[\left(\frac{1644 - 2501\bar{b}_k}{642}\right)\bar{y}'_{n+k+1} + \bar{b}_k\hat{y}'_{n+k} + \left(\frac{12 - 37\bar{b}_k}{214}\right)y'_{n+k-1}\right] \tag{6c}$$

$$\text{Corrector 2.}\quad \hat{y}_{n+k+1} - \frac{300}{137}\hat{y}_{n+k} + \frac{300}{137}y_{n+k-1} - \frac{200}{137}y_{n+k-2} + \frac{75}{137}y_{n+k-3}$$
$$-\frac{12}{137}y_{n+k-4} = h\left[\frac{12}{25}\hat{y}'_{n+k+1} - \frac{144}{3425}\bar{y}'_{n+k+1}\right] \tag{6d}$$

4. Concluding Remarks

As very briefly described in section 2 above, the construction of a good quality new block scheme is by no means easy. After considerable investigations, we arrived at a new block algorithm, which possesses certain good and essential stability characteristics (e.g. zero-stability). Unfortunately, due to space limitations, although there are several more important points to discuss, it is not possible to present a full account of the methods' properties.

References

[1]] H.A. Watts and L.F. Shampine: "A-stable block one-step methods", BIT vol. 12, pp. 252-266, 1972.

[2] J.E. Bond and J.R. Cash: "A Block method for the Numerical Integration of Stiff Systems of ODEs", BIT vol. 19, pp. 429-447, 1979.

[3] J.R. Cash, M.T. Diamantakis: "On the implementation of block Runge-Kutta methods for stiff IVPs", Ann. Numer. Math. 1, pp. 385-398, 1994.

[4] G. Psihoyios: "Advanced Step-point Methods for the solution of Initial Value Problems", PhD Thesis, Imperial College – University of London, 1995.

[5] G. Avdelas and T.E. Simos: "Block Runge-Kutta methods for periodic initial-value problems", Computers and Mathematics with Applications, vol. 31, pp. 69-83, 1996.

[6] L. Brugnano, D. Trigiante: "Solving Differential Problems by Multistep Initial and Boundary Value Methods", Gordon and Breach, London, 1998.

[7] F. Iavernaro, F. Mazzia: "Solving Ordinary Differential Equations by Generalized Adams methods: properties and implementation techniques", Applied Numerical Mathematics, vol. 28 (2-4), pp.107-126, 1998.

[8] L. Brugnano, D. Trigiante: "Block implicit methods for ODEs, Recent trends in Numerical Analysis", Nova Science Publishers Inc., Commack, NY, 2000.

[9] J.D. Lambert: "Numerical Methods for Ordinary Differential Systems", Wiley, Chichester 1991.

[10] J.R. Cash: "The Integration of Initial Value Problems in ODEs using Modified Extended BDF", Computers and Mathematics with Applications, vol. 9, pp. 645-657, 1983.

Brill Academic Publishers
P.O. Box 9000, 2300 PA Leiden,
The Netherlands

Lecture Series on Computer and Computational Sciences
Volume 2, 2005, pp. 102-105

Differential Entropy Approximations and Diffusive Restoration of Digital Images

F. Rodenas*[1], P. Mayo, D. Ginestar* and G. Verdú****

* Departamento de Matemática Aplicada,
** Departmento de Ingeniería Química y Nuclear,
Universidad Politécnica de Valencia, 46022-Valencia, Spain

Received 10 April, 2005; accepted in revised form 25 April, 2005

Abstract: One method successfully employed to denoise digital images is the diffusive iterative filtering. An important point of this technique is the estimation of the stopping time of the diffusion process. In this work, a stopping time criterion of the diffusive iterative filtering is proposed. It is based on the evolution of the differential entropy with the diffusion parameter. Because of computational complexity of the entropy function, it is estimated by using an approximation of the entropy function.

Keywords: Differential entropy, image restoration, diffusive filtering, stopping time.

1 Introduction

An important class of image denoising methods is based on nonlinear diffusion equations (see, for instance, [1, 2]). These methods are implemented as an iterative filter that has to be applied a number of steps. The final time or number of steps for the diffusive filter depends on the kind of noisy image to be restored and, usually, is selected attending to heuristic criteria. In this work we will evaluate the possibility of using the differential entropy function to develop an automatic stopping criterion for the image restoration process. Assuming that the noise present in the image is gaussian, while the diffusive filtering reduces the noise the image statistics become less and less gaussian. Therefore, is reasonable to think that a measure of the nongaussianity of the image can be used as a stopping condition of the algorithms. The possibility of using generalized entropies to choose stopping times in nonlinear diffusion scale-spaces was pointed by Sporring and Weickert in ref. [3].

2 Differential entropy approximations

The classical measure of nongaussianity of a random variable is kurtosis. However, it can be very sensitive to outliers. Thus, other measures of nongaussianity might be better than kurtosis. Another important measure of nongaussianity of a random variable is given by the differential entropy. The differential entropy H of a random variable X with probability density function $p(x)$ is defined as:

$$H(X) = -\int p(x)\, \log p(x)\, dx \,. \tag{1}$$

[1]Corresponding author. E-mail: frodenas@mat.upv.es

A fundamental well-known result of information theory is that a gaussian variable has the largest entropy among all random variables of equal variance. Thus, entropy could be used as a measure of nongaussianity.

Another useful concept is the negentropy function, which is defined in terms of the differential entropy H by:

$$J(X) = H(X_G) - H(X)\,, \tag{2}$$

where X_G is a Gaussian random variable of the same covariance matrix as X. $J(X)$ is always non-negative and it is zero if and only if X is a gaussian random variable.

Unfortunately, the estimation of differential entropy, or, equivalently, negentropy, using the definition is quite difficult because it requires the estimation of the density of X. This computational difficulty is overtaken by using simple and robust approximations of entropy. Useful approximations of entropy have been proposed by A. Hyvärinen in the context of independent component analysis [4]. These approximations are based on the estimation of the maximum entropy compatible with the measurements of the random variable X. Detailed derivations of the entropy approximations can be found in [4].

In the simplest case, the approximation of the negentropy function becomes:

$$J(X) \propto [\,E\{G(x)\} - E\{G(x_G)\}\,]^2\,, \tag{3}$$

where X and X_G are normalized to zero mean and unit variance, and G is a nonquadratic function. The more common choices for G that have proved very useful [4] are the following

$$G_1(x) = \frac{1}{a}\log\cosh(ax) \quad (1 \le a \le 2)\,, \quad G_2(x) = -\exp(-x^2/2)\,. \tag{4}$$

3 Image restoration method

3.1 Nonlinear diffusive filter with a constraint

The usual model considered for noisy images is the following: Let f be the observed image and let us assume that f is the sum of an ideal noise-free image $\bar{f}$ and a noise signal n:

$$f = \bar{f} + n\,. \tag{5}$$

We assume that the image $\bar{f}$ and the noise n are uncorrelated and the noise is gaussian and it has zero mean value. The initial point of the diffusion process is the observed image $f = u(0)$ and the iterative filtering produces a family of images $u(t)$ $(t = 0, 1, 2...)$. These images are filtered versions of the original f.

The nonlinear diffusive filter used in this paper is based on the dynamic equation (see [5] for more details) with a constraint:

$$\frac{\partial u}{\partial t} = \vec{\nabla}\left(\frac{\vec{\nabla}u}{\sqrt{\beta^2 + \left\|\vec{\nabla}u\right\|^2}}\right) + \epsilon\nabla^2 u + \mu(f - u)\,, \qquad \int_\Omega (u - f)^2\, d\vec{x} = \sigma^2 \int_\Omega d\vec{x}\,, \tag{6}$$

where β, μ and ϵ are constant, σ^2 is a estimation of the variance of the initial noise and Ω is a convex region of $\mathbb{R}^2$ constituting the support space of the surface u, representing the image.

The corresponding diffusive filter is obtained by discretizing both, spatially and temporally, the first equation in (6). We discretize the diffusion operator in each spatial node and for the time discretization we use an additive operator splitting (AOS) scheme [2], [5]. The values of the parameters for the filter used in this work are: $\epsilon = 0.1$, $\beta = 1$, $\mu = 0.1$ and time step $\Delta t = 0.1$.

3.2 Stopping time selection

An ideal diffusion filter which works optimally for a denoising task would first eliminate the noise before significantly affecting the signal, then, it would be possible to choose an image $u(T)$ from the diffusive family which, in some sense, is the nearest to the noise free signal $\tilde{f}$. In other words, for such a denoising filter, one should choose the stopping time T such that the restored image $u(T)$ is as near to the free-noise image $\tilde{f}$ as possible, this means that the euclidean distance $\|u(T) - \tilde{f}\|$ is as small as possible. Let us call such T that minimizes the distance $\|u(T) - \tilde{f}\|$ the optimal stopping time, T_{OPT}.

We want to select the stopping time which will be a good estimates of this optimal stopping time. The proposed criterion is based on the negentropy of the difference $(f - u(t))$. Let us consider the series $(f - u(t))$, $(t = 0, 1, 2...)$, then it is reasonable that the stopping time maximizes the gaussianity of the $(f - u(t))$. Using the negentropy as the measure of the gaussianity of a signal, this means that the stopping time can be chosen as the time that minimizes the negentropy of the 'noise' $(f - u(t))$. Then, the definition of the stopping time based on negentropy is

$$T_{NEG} = \arg\min J(f - u(t)) \,, \quad t = 0, 1, 2... \tag{7}$$

The main advantage of negentropy based method is that the diffusion stopping time obtained is estimated without any additional knowledge of the ideal free-noise image and using quite mild assumptions about noise. Computationally, the method is easily implemented since the negentropy $J(f - u(t))$ is estimated using the approximation (3) for the function G_1 in (4) with $a = 1.5$.

4 Experiments

The possibility of using the negentropy criterion to select a stopping time in the diffusion process is evaluated. To assess the applicability of this method, we take test images that we assume to be noise-free and corrupt them by adding different levels of gaussian noise.

Let us consider the light-house test image to show an example of the performance of the negentropy criterion. In Fig. 1(b), we observe the evolution of the negentropy of the difference $(f - u(t))$ and of the distance $\|u(t) - \tilde{f}\|$ with the diffusion time parameter for a noisy version of the light-house image. The minimum of the graph of $J(f - u(t))$ gives the negentropy-based

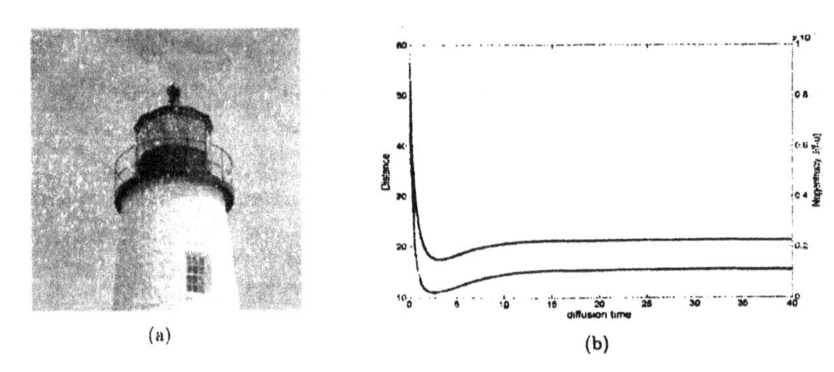

Figure 1: (a) Light-house test image. (b) The distance $\|u(t) - \tilde{f}\|$ and the negentropy $J(f - u(t))$ evolution with the diffusion time.

stopping time, T_{NEG}. In this example, we observe that the stopping time T_{NEG} is very similar to the optimal stopping time, T_{OPT}, which minimizes the euclidean distance $\|u(t) - \tilde{f}\|$. Using the criterion (7), we estimate stopping times for the test images corrupted with different levels of gaussian noise. The results are given in the following table, where the variance of the noise is given as a factor of the variance of the original image.

Table 1: Stopping times and distances for the light-house image.

σ^2 of noise	T_{NEG}	T_{OPT}	$\|u(T_{NEG}) - \tilde{f}\|$	$\|u(T_{OPT}) - \tilde{f}\|$
0.1	1.1	1	7.11	7.07
0.2	2.4	3.2	17.74	17.35
0.3	3.6	6.2	30.61	28.05
0.4	3.8	9.8	53.33	37.82

5 Conclusions

We have presented a new method to estimate a stopping time for iterative nonlinear diffusion filtering for image restoration. The stopping time selection is based on the idea that the negentropy of differences $f - u(t)$, which are estimations of the noise signal, must be minimized at stopping time. The negentropy is easily computed by using the approximations introduced in [4].

The experiments suggest that the negentropy criterion provides a good estimates of the stopping time in the diffusion process for moderate noise levels.

The results for the negentropy-based stopping time have been compared with the stopping time proposed by Weickert in [6], based on the notion of relative variance, and with the stopping time proposed by Mrázek and Navara in [7], based on the decorrelation criterion. As a result, we observe that the corresponding distances using the Mrázek and Navara method and the negentropy method are comparable with the optimal distances, but the stopping times of the diffusion process are lower using the negentropy criterion.

References

[1] F. Catté, P. Lions, J. Morel and T. Coll, Image Selective Smoothing and Edge Detection by Nonlinear Diffusion. *SIAM Numerical Analysis* **29** (1992), 182-193.

[2] J. Weickert, B.M. ter Haar Romeny, M.A. Viergever. Efficient and reliable schemes for nonlinear diffusion filtering, *IEEE Transactions on Image Processing* **7**(3), (1998), 398-410.

[3] J. Sporring and J. Weickert. Information measures in scale spaces. *IEEE Transactions on Information THeory* **45**(3) (1999), 1051-1058.

[4] A. Hyvärinen, New approximations of differential entropy for independent component analysis and projection pursuit. *Advances in Neural Information Processing Systems* **10** (1998), 273-279.

[5] J. Weickert. Efficient image segmentation using partial differential equations and morphology. *Pattern Recognition* **34**(9) (2001), 1813-1824.

[6] J. Weickert. Coherence-enhancing diffusion of colour images. *Image and Vision Computing* **17**(3-4) (1999), 201-212.

[7] P. Mrázek and M. Navara, Selection of optimal stopping time for nonlinear diffusion filtering. *International Journal of Computing Vision* **52**(2-3) (2003), 189-203.

Brill Academic Publishers
P.O. Box 9000, 2300 PA Leiden,
The Netherlands

Lecture Series on Computer and Computational Sciences
Volume 2, 2005, pp. 106-109

Asymptotic Complexity of Algorithms via the Nonsymmetric Hausdorff Distance

J. Rodríguez-López[1], S. Romaguera[1] and O. Valero[2,*]

[1] E.T.S. Ingenieros de Caminos, Canales y Puertos,
Departamento de Matemática Aplicada,
Universidad Politécnica de Valencia,
46071 Valencia, Spain
jrlopez@mat.upv.es; sromague@mat.upv.es

[2] Departamento de Matemáticas e Informática,
Universidad de las Islas Baleares,
07122 Palma de Mallorca, Baleares, Spain.
o.valero@uib.es

Received 7 April, 2005; accepted in revised form 25 April, 2005

Abstract: The theory of nonsymmetric topology has been successfully applied to the theory of complexity of algorithms. In 1995, Schellekens introduced a quasi-pseudo-metric space in order to obtain a mathematical model for the study of complexity of algorithms. Here, we provide a mathematical context to study the asymptotic complexity of algorithms.

Keywords: asymptotic complexity, algorithms, Hausdorff quasi-metric.

Mathematics Subject Classification: 68Q25, 54E35, 54B20.

1 Introduction

One of the main interesting problems which arise in algorithm analysis is the study of how much memory or how much execution time an algorithm needs in order to obtain a result from a certain entry. This process is influenced by a lot of factors like the software or technological features. In this way, it then appears the concept of asymptotic analysis and complexity.

In 1995, Schellekens [8] began the development of a mathematical model to study the complexity analysis of algorithms. In this direction, he found useful the theory of nonsymmetric topology. We recall that a quasi-pseudo-metric space is a pair (X, d) such that X is a nonempty set and d is a nonnegative real-valued function defined on $X \times X$ such that for all $x, y, z \in X$:

i) $d(x, x) = 0$

ii) $d(x, z) \leq d(x, y) + d(y, z)$

The function d is then called a *quasi-pseudo-metric on X*. If d also verifies the condition (i') $d(x, y) = 0$ if and only if $x = y$, then d is said to be a *quasi-metric*. Furthermore, if d is a symmetric function, i. e. $d(y, x) = d(x, y)$ we say that d is a *metric*.

* The authors thank the support of the Plan Nacional I+D+I and FEDER, grant BFM2003-02302.

Thus, Schellekens introduced and discussed the so-called *complexity* (quasi-pseudo-metric) *space* which is the pair $(\mathcal{C}, d_\mathcal{C})$, where $\mathcal{C}$ is the set of complexity functions, i.e. $\mathcal{C}$ consists of all functions f from the set ω of all nonnegative integer numbers into $(0, \infty]$ such that

$$\sum_{n=0}^{\infty} 2^{-n} \frac{1}{f(n)} < \infty$$

and $d_\mathcal{C}$ is the quasi-pseudo-metric on $\mathcal{C}$ given by

$$d_\mathcal{C}(f, g) = \sum_{n=0}^{\infty} 2^{-n} \left[\left(\frac{1}{g(n)} - \frac{1}{f(n)} \right) \vee 0 \right], \qquad f, g \in \mathcal{C}.$$

In this way, each algorithm is represented by its complexity function. This space is useful in the theory of Computer Science and it is interesting from a theoretical and practical point of view. In fact, Schellekens used the Banach fixed point theorem in order to obtain an optimization result about Divide & Conquer algorithms (see [8]).

Several authors have studied this space and its applications (see [2, 3, 4, 5, 6, 7]). A modification of the complexity space, called the *dual complexity space*, was introduced by Romaguera and Schellekens in [5] because it provides a more appropriate mathematical context. By means of this concept, several quasi-metric and topological properties of the complexity space have been discussed. In fact, they proved that the dual complexity space is isometric to the complexity space and it has good properties like Smyth completeness which provides a consistent setting in Domain Theory, in particular in denotational semantics of programming languages (see [9]).

In [2] the notion of dual complexity space was extended to the "p-dual" case, motivated by the fact that several algorithms which are exponential time complexity do not provide complexity functions in $\mathcal{C}^*$. For instance, an algorithm with running time $\mathcal{O}(2^n/\sqrt{n})$ corresponds to the complexity function f, given by $f(n) = 2^n/\sqrt{n}$ that obviously does not belong to the dual complexity space. The p-dual space is constructed in the following way. For $p \in [1, +\infty)$, the *dual p-complexity space* is the quasi-pseudo-metric space $(\mathcal{C}_p^*, d_p)$ where $\mathcal{C}_p^* = \{f : \omega \to [0, \infty) : \sum_{n=0}^{\infty}(2^{-n} f(n))^p < +\infty\}$ and

$$d_p(f, g) = \left(\sum_{n=0}^{\infty} (2^{-n}((g(n) - f(n)) \vee 0))^p \right)^{1/p}, \qquad f, g \in \mathcal{C}_p^*.$$

Therefore, the distance $d_p(f, g)$ defined in the p-dual complexity space $\mathcal{C}_p^*$ between two complexity functions f and g measures the progress made in lowering the complexity by replacing any program P with complexity function f by any program Q with complexity function g. If $d_p(f, g) = 0$ we can deduce that g is more efficient than f.

In [4] a new (extended) quasi-metric e_p is defined in the dual p-complexity space, which has better topological properties than the classical one. e_p is the following invariant balanced extended quasi-metric defined on $\mathcal{C}_p^*$:

$$e_p(f, g) = \begin{cases} d_p(f, g) & \text{if } f \leq g \\ +\infty & \text{otherwise} \end{cases}.$$

On the other hand, it is usual in the study of complexity to work with asymptotic notation. This process consists in studying the behavior of the complexity of the algorithm when the input grows infinitely. In this way, algorithms are grouped in classes which have the same asymptotic behavior. In this paper, we provide a new approach to the problem of obtaining an appropriate mathematical model for the study of algorithms using the asymptotic complexity theory. We will do it by means of the Hausdorff quasi-pseudo-metric which is one of the main techniques to endow with a quasi-pseudo-metric the family of closed sets of a quasi-pseudo-metric space.

2 Asymptotic complexity

The asymptotic complexity of an algorithm provides a characterization of its efficiency and studies the behavior of the algorithm when a big input is considered.

Three kinds of asymptotic complexity can be studied (see [1]):

- $\mathcal{O}(f) = \{g \in \mathcal{C}_p^* :$ there exist $c > 0$ and $k_0 \in \mathbb{N}$ such that $g(k) \leq cf(k)$ for all $k \geq k_0\}$.
- $\Omega(f) = \{g \in \mathcal{C}_p^* :$ there exist $c > 0$ and $k_0 \in \mathbb{N}$ such that $cf(k) \leq g(k)$ for all $k \geq k_0\}$.
- $\Theta(f) = \mathcal{O}(f) \cap \Omega(f)$.

The quasi-metric e_p introduced in [4] provides a suitable measurement of the distance between f and g if $f \leq g$. Therefore, in this case $f \in \mathcal{O}(g)$, so we can only obtain measures in the same class. For example, if $f(k) = 4k$ and $g(k) = 2^k/k$ then $e_p(f,g) = +\infty$, for every $p \geq 1$, but it seems to be clear that f is much more efficient than g. So, we think that a slight modification of this definition could provide a more efficient tool for the study of asymptotic complexity. We will use the following notation: $f \leq_* g$ if there exists $k_0 \in \mathbb{N}$ such that $f(k) \leq g(k)$ for all $k \geq k_0$.

The following result is similar to [4, Theorem 1].

Proposition 1 *For each $p \in [1, +\infty)$, we consider the function $e'_p : \mathcal{C}_p^* \times \mathcal{C}_p^* \to [0, +\infty]$ defined by*

$$e'_p(f,g) = \begin{cases} d_p(f,g) & \text{if } f \leq_* g \\ +\infty & \text{otherwise} \end{cases}.$$

Then e'_p is an invariant extended quasi-pseudo-metric on $\mathcal{C}_p^*$.

Lemma 1

1. *$\mathcal{O}(f)$ is $\mathcal{T}(e'_p)$-closed for all $p \in [1, +\infty)$.*
2. *$\Omega(f)$ is $\mathcal{T}((e'_p)^{-1})$-closed for all $p \in [1, +\infty)$.*
3. *$\Theta(f)$ is $\mathcal{T}((e'_p)^s)$-closed or all $p \in [1, +\infty)$.*

As a consequence of the above result we give the following definition.

Definition 1 *Let us define $\mathcal{O}_p = \{\mathcal{O}(f) : f \in \mathcal{C}_p^*\}$ and $\Omega_p = \{\Omega(f) : f \in \mathcal{C}_p^*\}$. The **p-upper complexity space** (resp. p-lower complexity space) is the quasi-pseudo-metric space $(\mathcal{O}_p, H_p^-)$ (resp. (Ω_p, H_{-p}^+)), where H_p^- (resp. H_{-p}^+) denotes the lower (resp. upper) Hausdorff quasi-pseudo-metric induced by e'_p (resp. $(e'_p)^{-1}$).*

*The **p-complexity space** is the pseudo-metric space (Θ_p, H_p) where $\Theta_p = \mathcal{O}_p \cap \Omega_p$ and $H_p = (H_p^-)^s = \max\{H_p^-, (H_p^-)^{-1}\} = \max\{H_p^-, H_{-p}^+\}$.*

This new space provides an appropriate context to study complexity of algorithms. In fact, if $e_p(f,g) = 0$ then f is more efficient than g (see the Introduction). It is easy to prove that $f \in \mathcal{O}(g)$ if and only if $\mathcal{O}(f) \subseteq \mathcal{O}(g)$, and this is also equivalent to $H_p^-(\mathcal{O}(f), \mathcal{O}(g)) = 0$. Therefore, the quasi-pseudo-metric H_p^- keeps the meaning of e_p. A similar reasoning could be made with the p-lower complexity space.

We will study some interesting properties of the asymptotic complexity space. For instance, we can prove the following result.

Proposition 2

1. *The quasi-pseudo-metric H_p^- is subinvariant.*

2. $(\mathcal{O}_p, H_p^-)$ *is right K-complete.*

3. $(\mathcal{O}_p, H_p^-)$ *is compact.*

3 Algebraic properties of the asymptotic complexity space

In the study of complexity of algorithms it is usual to make algebraic manipulations with complexity functions. For example, to obtain the running time of an algorithm we sum the running time of each sentence, and if the algorithm contains a boucle we multiply the running time of the boucle by the number of repetitions. Algebraic operations also appear when computing the running time of a recursive algorithm, like a recursion tree (see [1]).

In this work we also discuss some algebraic properties of the asymptotic complexity space. For example, we obtain the following result.

Proposition 3 *The p-upper complexity space* $(\mathcal{O}_p, H_p^-)$ *is a topological abelian monoid with the pointwise sum of functions.*

References

[1] T. H. Cormen, C. E. Leiserson, R. L. Rivest and C. Stein, *Introduction to algorithms*, MIT Press, 2001.

[2] L. M. García-Raffi, S. Romaguera and E. A. Sánchez-Pérez, *Sequence spaces and asymmetric norms in the theory of computational complexity*, Math. Comput. Model. **36** (2002), 1–11.

[3] L. M. García-Raffi, S. Romaguera and E. A. Sánchez-Pérez, *The dual space of an asymmetric normed linear space*, Quaest. Math. **26** (2003), 83–96.

[4] S. Romaguera, E. A. Sánchez-Pérez and O. Valero, *Computing complexity distances between algorithms*, Kybernetika **39** (2003), 569–582.

[5] S. Romaguera and M. Schellekens, *Quasi-metric properties of complexity spaces*, Topology Appl. **98** (1999), 311–322.

[6] S. Romaguera and M. Schellekens, *The quasi-metric of complexity convergence*, Quaest. Math. **23** (2000), 359–374.

[7] S. Romaguera and M. Schellekens, *Duality and quasi-normability for complexity spaces*, Appl. Gen. Topology **3** (2002), 91–112.

[8] M. Schellekens, *The Smyth completion: A common foundation for denotational semantics and complexity analysis*, Proc. MFPS 11, Electron. Notes Theor. Comput. Sci. **1** (1995), 211–232.

[9] M. B. Smyth, *Quasi-uniformities: Reconciling domains with metric spaces*, Lect. Notes Comput. Sci. **298** (1988), 236–253.

Brill Academic Publishers
P.O. Box 9000, 2300 PA Leiden,
The Netherlands

Lecture Series on Computer and Computational Sciences
Volume 2, 2005, pp. 110-115

Numerical Computation and Experimental Study of The Stress Field of En-Echelon Cracks

GUO Rongxin[a†] LI Junchang[b] Xiong Bingheng[b]

[a]Department of Engineering Mechanics, Kunming University of Science and technology, 253 Xuefu Road, Kunming 650093,China
[b]Department of Physics, Kunming University of Science and technology, 253 Xuefu Road, Kunming 650093,China

Received 9 April, 2005; accepted in revised form 26 April, 2005

Abstract: The PMMA plate is used to simulate the brittle material and en echelon cracks are prefabricated on the plates. The computer program dealing with the interaction among multi inclusions has been written based on the equivalent inclusion method modified from Moschovidis' procedure. The stress field of En-echelon Cracks has been computed by the program and compared well with those observed by real-time holographic interferometry. The dark shadow areas observed in the real time hologram experiment are the phenomena of caustic in Geometry Optics. The stress intensity factor can be calculated by measuring the size of the corresponding length of the shadow spot according to Fracture Mechanics theory.

Keywords: Equivalent inclusion method, Interaction, Stress field, Numerical computation, real-time holographic interferometry.

Mathematics SubjectClassification: 74M25

PACS: 46.50.+a

1. Introduction

The En-echelon Cracks widely exit in brittle materials, especially in rock. It plays a very important role in the fracture process of rock. The failure mechanism of rock has been studied for many years with the help of mesomechanics. The equivalent inclusion method proposed by J. D. Eshelby is one of the pioneer work of mesomechanics[1,2]. Moschovidis investigated the applicability of the equivalent inclusion method to obtain numerical solutions for one or two inhomogeneities and pressurized cavities of any shape[3]. In the Moschovidis' procedure, the eigenstrain, applied field and the induced field were expressed in the form of polynomials of coordinates, and the derivations were specialised for the ellipsoidal inclusion with the eigenstrain given as the polynomial. Some of the inclusion and inhomogeneity problems may be solved for a region of arbitrary shape. An elliptical crack is a special case of an ellipsoidal void where one of the principal axes of the ellipsoid becomes vanishingly small, therefore cracks and inclusions including the transformation toughening, crack growth through composites and stress intensity factors have been investigated by the equivalent inclusion method.[4]

The computer program dealing with the interaction among multi inclusions has been written based on the equivalent inclusion method modified from Moschovidis' procedure.[5] This program can be run under Windows environment with the help of Delphi 7.0 and the calculation results can be illustrated by Matlab. The stress fields of one and two cavities under uniaxial applied stress are computed by this program and compared well with the exact results as well as numerical computation results in the literatures. The special ellipsoidal cavity is used to simulate crack, and then this computer program is used to calculate the stress field of En-echelon Cracks in this paper.

2. The elastic fields of the collective inhomogeneities with an eigenstrain given as a polynomial

The equivalent eigenstrains have to be assumed as polynomials of degree greater than or equal to the degree of the applied strain polynomial; the higher the degree of the eigenstrains, the better the

† Corresponding author. E-mail: guorx@kmust.edu.cn, grx_99@yahoo.com

accuracy of the solution. We have compared the computation results of assuming the polynomial to be different degree. It is obvious that the computation results are accurate enough and the formula is not very complex when the eigenstrains are assumed to be the polynomials of degree two.

It should be emphasized that ε_{kl} is no longer uniform even if σ_{ij}^{a} is uniform, since interior points of one inhomogeneity are exterior points of other inhomogeneities and the stress in this inhomogeneity is disturbed by other inhomogeneities.

Let $\mathbf{OX_1X_2X_3}$, $\overline{\mathbf{OX_1X_2X_3}}$, and $\overline{\overline{O\,\mathbf{X_1X_2X_3}}}$ be the coordinate systems in the matrix and the inhomogeneities be ellipsoids with semiaxes coinciding with the axes of the coordinate systems. From Moschovidis' procedure, the applied field, the equivalent eigenstrains and the strain fields of every inhomogeneitiy, $\varepsilon_{ij}^{a}(x)\ \varepsilon_{ij}^{a}(\bar{x})$, $\varepsilon_{ij}^{*I}\ \varepsilon_{ij}^{*II}$ can be given by following polynomials:

$$\varepsilon_{ij}^{a}(x) = E_{ij} + E_{ijk}x_k + E_{ijkl}x_kx_l + \cdots$$

$$\overline{\varepsilon}_{ij}^{a}(x) = \overline{E}_{ij} + \overline{E}_{ijk}\bar{x}_k + \overline{E}_{ijkl}\bar{x}_k\bar{x}_l + \cdots$$

$$\varepsilon_{ij}^{*I}(x) = B_{ij}^{I} + B_{ijk}^{I}x_k + B_{ijkl}^{I}x_kx_l + \cdots$$

$$\overline{\varepsilon}_{ij}^{*II}(x) = B_{ij}^{II} + B_{ijk}^{II}\bar{x}_k + B_{ijkl}^{II}\bar{x}_k\bar{x}_l + \cdots$$

$$\varepsilon_{ij}^{I}(x) = D_{ijkl}^{I}(x)B_{kl}^{I} + D_{ijklq}^{I}(x)B_{klq}^{I} + D_{ijklqr}^{I}(x)B_{klqr}^{I} + \cdots$$

$$\varepsilon_{ij}^{II}(x) = D_{ijkl}^{II}(\bar{x})B_{kl}^{II} + D_{ijklq}^{II}(\bar{x})B_{klq}^{II} + D_{ijklqr}^{I}(\bar{x})B_{klqr}^{II} + \cdots$$

Where I, II are the serial number of inhomogeneities and $x_i, \bar{x}_i\bar{\bar{x}}_i$ are the coordinate systems in the matrix corresponding to each inhomogeneity.
The relation between two coordinate systems is given as

$$x_i - c_i = a_{ij}\bar{x}_j \quad \text{and} \quad \bar{x}_i = a_{ji}(x_j - c_j)$$

where a_{ij} is the direction cosine between the x_i-axis and the $\bar{x}_j$-axis, and c_i is the x_i-coordinate of the origin $\bar{o}$ in Ω_2.
Because the ε_{ij}^{*} is given in the form of polynomial, the induced field $\varepsilon_{ij}(x)$ should also be expressed in form of polynomial in x_i. This is achieved by expanding $\varepsilon_{ij}(x)$ in Taylor's series around the origin o of the corresponding coordinate system. So the strain at point P can be approximately expressed as

$$\varepsilon_{mn}[p] = \varepsilon_{mn}[0] + \varepsilon_{mn},p[0]x_p^{P} + \frac{1}{2}\varepsilon_{mn,pq}[0]x_p^{P}x_q^{P} + \cdots$$

With the help of the superposition principle of the elasticity theory, the strain induced by the eigenstrain in every inhomogeneity at point P is

$$\varepsilon_{ij}(p) = \varepsilon_{ij}^{I}(p) + \varepsilon_{ij}^{II}(p) + \varepsilon_{ij}^{III}(p)$$

Taking the interaction among the ellipsoidal inhomogeneities into consideration, and the coefficients of the power series in the left- and right-hand sides in the equivalent equation are equated. We have the algebraic equations for each inclusion in its corresponding coordinate as following:
InΩ_1,

$$\Delta C_{stmn}^{I}\{[D_{mnij}^{I}[0]B_{ij}^{I} + D_{mnijk}^{I}[0]B_{ijk}^{I} + D_{mnijkl}^{I}[0]B_{ijkl}^{I} + \cdots] + a_{mc}a_{nh}[D_{chij}^{II}[0]B_{ij}^{II} +$$

$$D_{chijk}^{II}[0]B_{ijk}^{II} + D_{chijkl}^{II}[0]B_{ijkl}^{II} + \cdots]\} - C_{stmn}B_{mn}^{I} = -\Delta C_{stmn}^{I}E_{mn}$$

$$\Delta C_{stmn}^{I}\{[\frac{\partial}{\partial x_p}D_{mnij}^{I}[0]B_{ij}^{I} + \frac{\partial}{\partial x_p}D_{mnijk}^{I}[0]B_{ijk}^{I} + \frac{\partial}{\partial x_p}D_{mnijkl}^{I}[0]B_{ijkl}^{I} + \cdots] + a_{mc}a_{nh}a_{pf}$$

$$[\frac{\partial}{\partial \bar{x}_f}D_{chij}^{II}[0]B_{ij}^{II} + \frac{\partial}{\partial \bar{x}_f}D_{chijk}^{II}[0]B_{ijk}^{II} + \frac{\partial}{\partial \bar{x}_f}D_{chijkl}^{II}[0]B_{ijkl}^{II} + \cdots]\} - C_{stmn}B_{mnp}^{I} = -\Delta C_{stmn}^{I}E_{mnp}$$

$$\frac{1}{2!}\Delta C^{I}_{stmn}\{[\frac{\partial^2}{\partial x_p \partial x_q} D^{I}_{mnij}[0]B^{I}_{ij} + \frac{\partial^2}{\partial x_p \partial x_q} D^{I}_{mnijk}[0]B^{I}_{ijk} + \frac{\partial^2}{\partial x_p \partial x_q} D^{I}_{mnijkl}[0]B^{I}_{ijkl} + \cdots] +$$

$$a_{mc}a_{nh}a_{pf}a_{qg}[\frac{\partial^2}{\partial \bar{x}_f \partial \bar{x}_g} D^{II}_{chij}[0]B^{II}_{ij} + \frac{\partial^2}{\partial \bar{x}_f \partial \bar{x}_g} D^{II}_{chijk}[0]B^{II}_{ijk} + \frac{\partial^2}{\partial \bar{x}_f \partial \bar{x}_g} D^{II}_{chijkl}[0]B^{II}_{ijkl} + \cdots]\}$$

$$- C_{stmn}B^{I}_{mnpq} = -\Delta C^{I}_{stmn}E_{mnpq}$$

etc.

Similarly, we can have the algebraic equations in Ω_2.

The above algebraic equations can be solved approximately if the applied field is given as a polynomial of a finite degree, for any shape of the inhomogeneity. In our numerical computation program, the applied field is given as a polynomial of degree zero, and the eigenstrain is assumed as polynomials of degree two.

The orientation of each inhomogeneity is determined by three angles, θ φ and ψ .

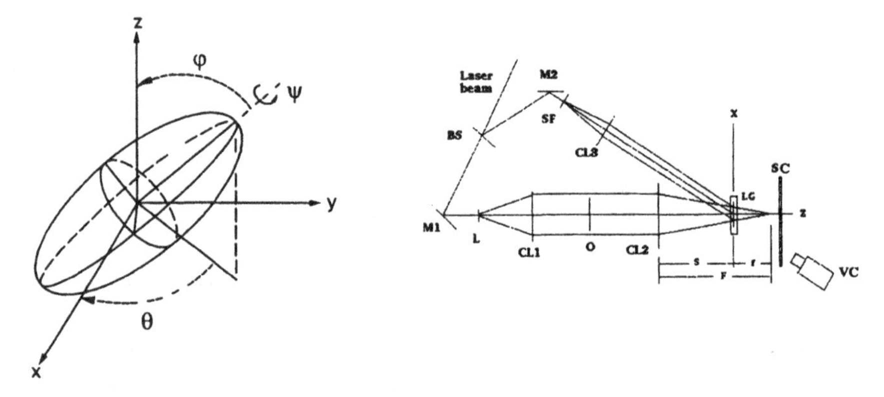

Fig. 1 The orientation of inhomogeneity

Fig. 2. Schematic diagram of the real-time holographic optical setup in our experiments
BS: Beam splitter M1, M2: Mirror SF: Spatial filter CL1: Φ 300mm Collimator O: Specimen VC: Video camera CL2: Φ 450mm Imaging lens CL3: Φ100mm Collimator L: Concave lens LG: Liquid gate SC observation screen

3. Numerical computation

Many of the Moschovidis' numerical computation results[3], illustrated in figure 3 to figure 7 and figure 37 to figure 43 as well as figure 48 and figure 49 in his dissertation, are compared with those of the results computed by this computer program. Computation results computed by this program are almost identical with those of Moschovidis' except those in figure 38(a) and figure 38(b). Besides, computation results computed by this computer program are also compared well with those of Arnaud Riccardi's[6] illustrated in figure16 to figure 18 and E. Sternberg's[7]. Based on the above conclusion, it can be said that this numerical computation program is reasonable

In order to simulate the En-echelon Cracks, the stress fields for two kinds of crack configuration have been calculated. The calculation results of the sum of principal stress is illustrated in figure 3 and figure 4. The uniaxial applied stress is compress and its direction is along y axis. Figure 3 (a) and figure 4(a) are the contour figure respectively.

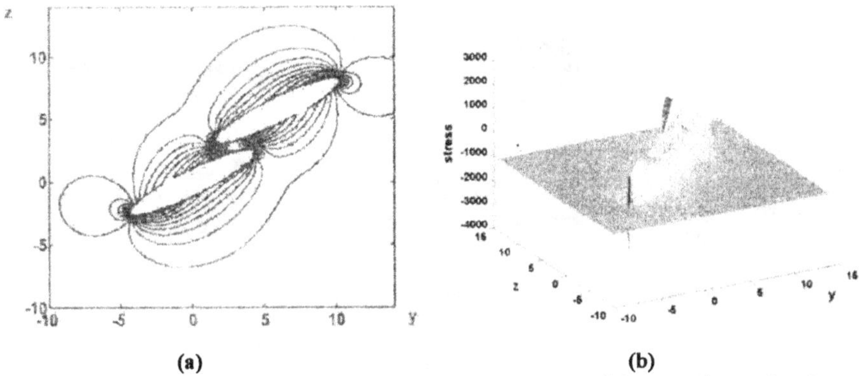

(a) **(b)**

Fig 3 the stress distribution of the sum of principal stress ($\sigma_{yy}+\sigma_{zz}$) for configuration 1

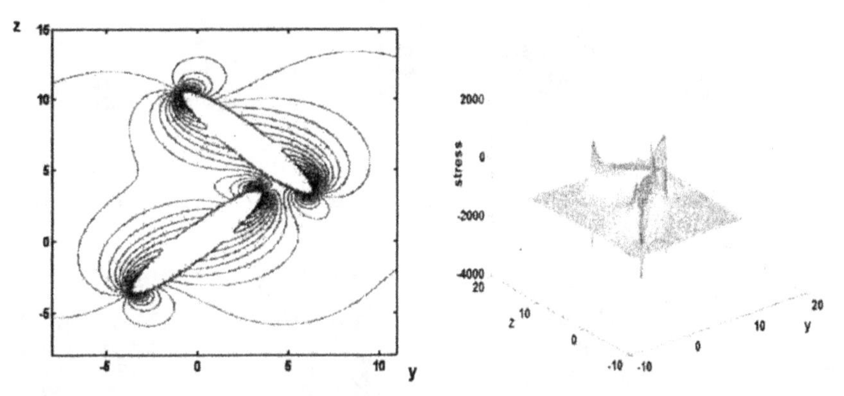

Fig 4 the stress distribution of the sum of principal stress ($\sigma_{yy}+\sigma_{zz}$) for configuration 2

4. Experiment

The PMMA plate is used to simulate the brittle material and en echelon cracks are prefabricated on the plates. The Stress field of En-echelon Cracks is observed by real-time holographic interferometry.

5. Conclusion

It can be seen that the distribution of stress intensity calculated from computation program compare well with those observed from experiment. The dark shadow areas observed in the real time hologram experiment are the phenomena of caustic in Geometry Optics. The stress intensity factor can be calculated by measuring the size of the corresponding length of the shadow spot according to Fracture Mechanics theory.

There exits a high tension stress region between two cracks and PMMA is a kind of brittle material. This is why that the specimen broke along z axis.

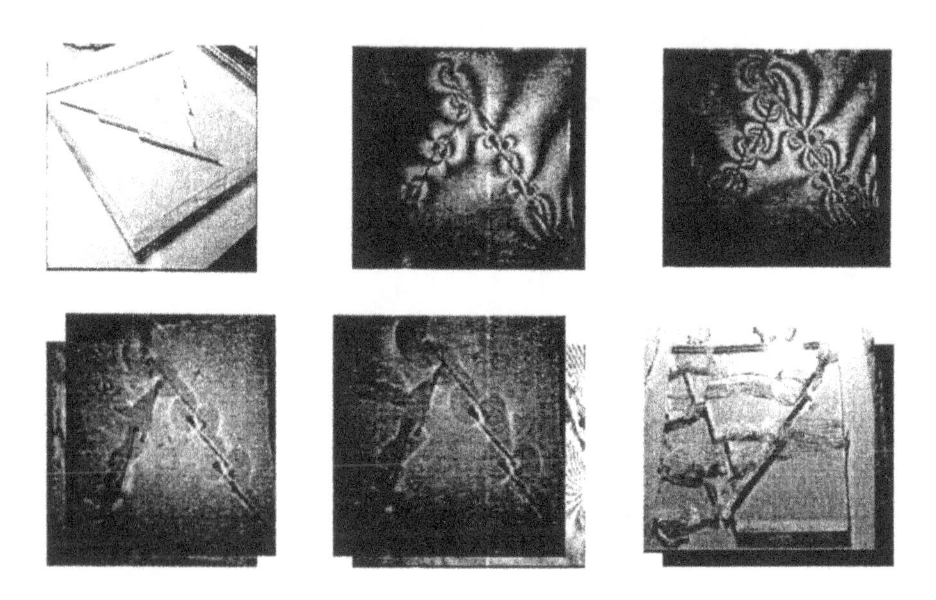

Figure 5 the distribution of stress intensity in whole fracture process of 1# specimen under compress load

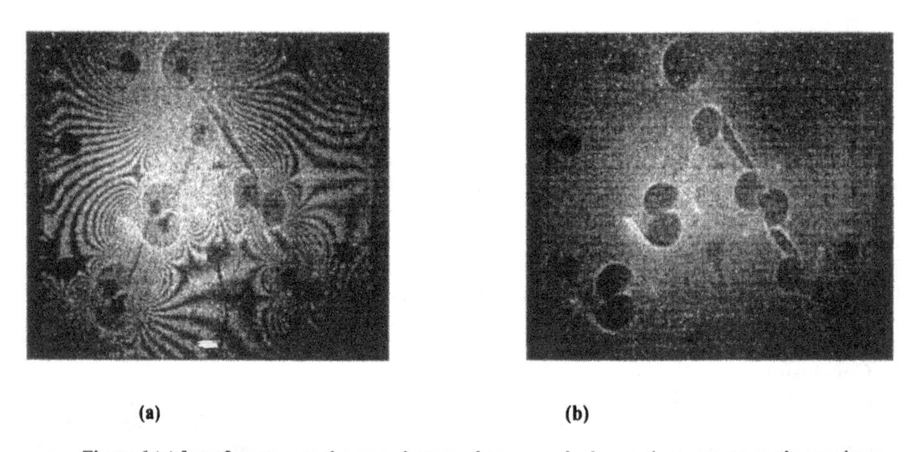

(a) (b)

Figure 6 (a) Interferogram at the second stage when some shadow regions appear on the specimen
(b) The corresponding photograph of (a) when the real time hologram illuminated by the object wave only

Acknowledgments

The authors would like to acknowledge the National Natural Science Foundation of China (10462002) and the Natural Science Foundation of Ynunan Province(2004A0011M) for their financial support and Gerard LORMAND, GEMPPM INSA de Lyon, for his help in modifying the program. The author also wishes to thank the anonymous referees for their careful reading of the manuscript and their fruitful comments and suggestions.

References

[1] J.D. Eshelby, *The determination of the elastic field of an ellipsoidal inclusion and related problems.* Proc. Roy. Soc. A, 1957, 241,376-396.

[2] J.D. Eshelby. *The elastic field outside an ellipsoidal inclusion.* Proc. Roy. Soc. A, 1959, 252, 561-569.

[3] Z.A. Moschovidis. *Two Ellipsoidal Inhomogeneities and Related Problems Treated by the Equivalent Inclusion Method.* Dissertation for the Ph.D, Northwestern University, 1975.

[4] T. Mura. *Micromechanics of Defects in Solids*, revised edition. Martinus Nijhoff Publishers,. 1987.

[5] R. X. GUO, G. LORMAND, J. C. LI. *Numerical Computation Program for the Inclusion Problem.* J. Kunming Univ. Sci. and Tech. 2004, 29 (3) 51-55 (In Chinese).

[6] Arnaud Riccardi, (1998) *Theoretic Study on the Nonclassical Inclusion Configuration in the Linear Elastic Isotrope*, Dissertation for the Ph.D., National Polytechnic Institute of Grenoble.

[7] E. Sternberg, M. A. Sadowsky, *On the Axisymmetric Problem of the Theory of Elasticity for an Infinite Region Containing Two Spherical Cavities.* J. Appl. Mech. pp19-27(1952).

Brill Academic Publishers
P.O. Box 9000, 2300 PA Leiden,
The Netherlands

Lecture Series on Computer and Computational Sciences
Volume 2, 2005, pp. 116-119

Modelling Business to Consumer Systems Adoption

J.L. Salmeron[1] and J.M. Hurtado

University Pablo de Olavide,
41013 Seville, Spain

Received 25 January, 2005; accepted in revised form 16 March, 2005

Abstract: For firms engaging in conventional "brick and mortar" activities, the use of EC can offer a new opportunity to extend and increase their market value. But, achieving success with B2C is difficult. To know with precision, the decision maker's expectations before implementation of the tool, can give important information for project development success. For this reason the authors consider this paper a useful endeavour that can help to identify the true reasons that decision makers have to establish B2C systems.

Keywords: Modelling, B2C, End-user computing, Cognitive maps

1. Introduction

The numbers of traditional businesses that are using Business to Consumer systems (B2C), as a new channel are growing quickly. The first objective is to use this for businesses advertising, but a great deal of them evolve adding other different services, like ordering channel or web customer support channel. For firms engaging in conventional "brick and mortar" activities, the use of EC can offer a new opportunity to extend and increase their market value (Subramani and Walden, 2001).
Achieving success with B2C is difficult. To know with precision, the decision maker's expectations before implementation of the tool, can give important information for project development success. For this reason the authors consider this paper a useful endeavour that can help to identify the true reasons that decision makers have to establish B2C systems. And next, establish technology strategy together in relationship with these reasons.

2. Theoretical background

Since the EC strategic decision is rarely planned some companies do not know what the main objectives are or what they actually want to achieve. Profitable positioning of EC necessitates a coherent articulation with both, the global strategic objectives and the business processes. B2C, and EC in general, can produce positive effects for corporations (Changa et al., 2003). Firms can expect both, offer more services and increase their sales by adding the web to their traditional channels. B2C can be used for attracting new customers, reaching new markets, creating new distribution channels and offering value-added customer services (Chatterjee and Grewal, 2002).
The success or failure of a B2C implementation may be determined by how well it identifies its true goals and clarify which competitive advantage it pursues. This can represent a vital step. Brick and mortar companies that incorporate a B2C strategy in corporate strategy should receive benefits from this fact but if the B2C is not considered at the top level, firms may not be so successful in their on-line results (Changa et al., 2003).
Non previous works have been found that modelling B2C adoption. This paper can provide important insight in the study of reasons why businesses operating in the fashion industry run their B2C ideas, and their applicability to other industries. Another important aspect, by which we have considered this

[1] Corresponding author. Pfr. Dr. Jose L. Salmeron. E-mail: jlsalsil@upo.es

industry, is because though almost all the main companies operating in the fashion industry have a Web site, but a few of them have B2C capabilities.

3. B2C modelling

A cognitive map gives a representation of how the people think about a particular issue or situation, by analyzing, arranging the problems and mapping graphically concepts that are connected between them. In addition, it identifies causes and effects and explains causal links (Eden and Ackermann, 1992). The cognitive maps study how people perceive the world around them and the way they act to reach their desires within their world. Kelly (1955; 1970) gives the foundation for this theory, based in a particular cognitive psychology body of knowledge. The base postulate for the theory is that "a person's processes are psychologically canalized by the ways in which he anticipates events".

We have chosen this technique because we consider that it is the most appropriate way to analyze the decision making process in relation to the adoption of B2C. Besides, since to the number of constructs that a Top Manager can use in the process of decision making is small (Calori et al., 1994), we think that mapping can also help to represent all the ideas and extract those which can be the most relevant in this process. It is a critical issue in real implementations.

The data eliciting process began with an interview in order to have a first approximation of the characteristic surrounding the company and the industry. Industry issues such as markets, competitors, and attributes were discussed.

Following, in the second session the CEO of the company, gave information demanded in the latest meeting and was informed about the subject that must be dealt with in the mapping session. Themes like the aim of the session, what is expected of him, and how the software operates were taken into consideration. Following a presentation of the consideration of the problem the CEO began to explore issues relating to the theme and its resolution.

In the third session, the mapping took place, and finally in the last session, the result was given to the participating members to verify the validity of the maps and facilitate theirs discussion.

4. Findings and Discussion

Findings suggest that among the most important reasons that are associated with the reason to establish B2C, increase of revenue, facilitate the purchases of consumers and amplify the brand recognition, hold the three most strongest positions.

There are 27 valid constructs considered in this study. The data indicates that 89 % of them exert positive influence in establish B2C.

In relation to the competitive advantage that each concept represents, 29,6 % corresponds to growth, 40,7 % differentiation and 18,5 % cost reduction. The result in percentage is changes if we study only the ten constructs with the highest density. In this case, in relation with competitive advantage, growth reaches 60% and differentiation 40%. Obviously, in this process flows the causal relationship with dominant and dominate concepts which make different, the importance of each one.

To increase the revenue holds the first position of the most important reasons not only for its own importance, but by the influence that the other concepts have on it. Follow the line of argument, amplify the brand recognition, increase of sales in other channels, gain other markets, enlarge the number of clients, facilitate the purchase of consumers, and augment the rotation of some articles, have a direct positive influence in this concept.

Like has been exposed, the main concept of two cited competitive advantages are highly interconnected. That can explain why the results of both tables are different. These facts appear to be the most important in the strategic decision process of establishing B2C.

Another aspect to considerer is that the companies' situation and managers' needs can influence the cause cognitive mapping. The hypothesized effect of cost reduction that B2C can give could not be considered so significant in the previous process of establishing B2C decision making if people are thinking for instance, in strategic growth.

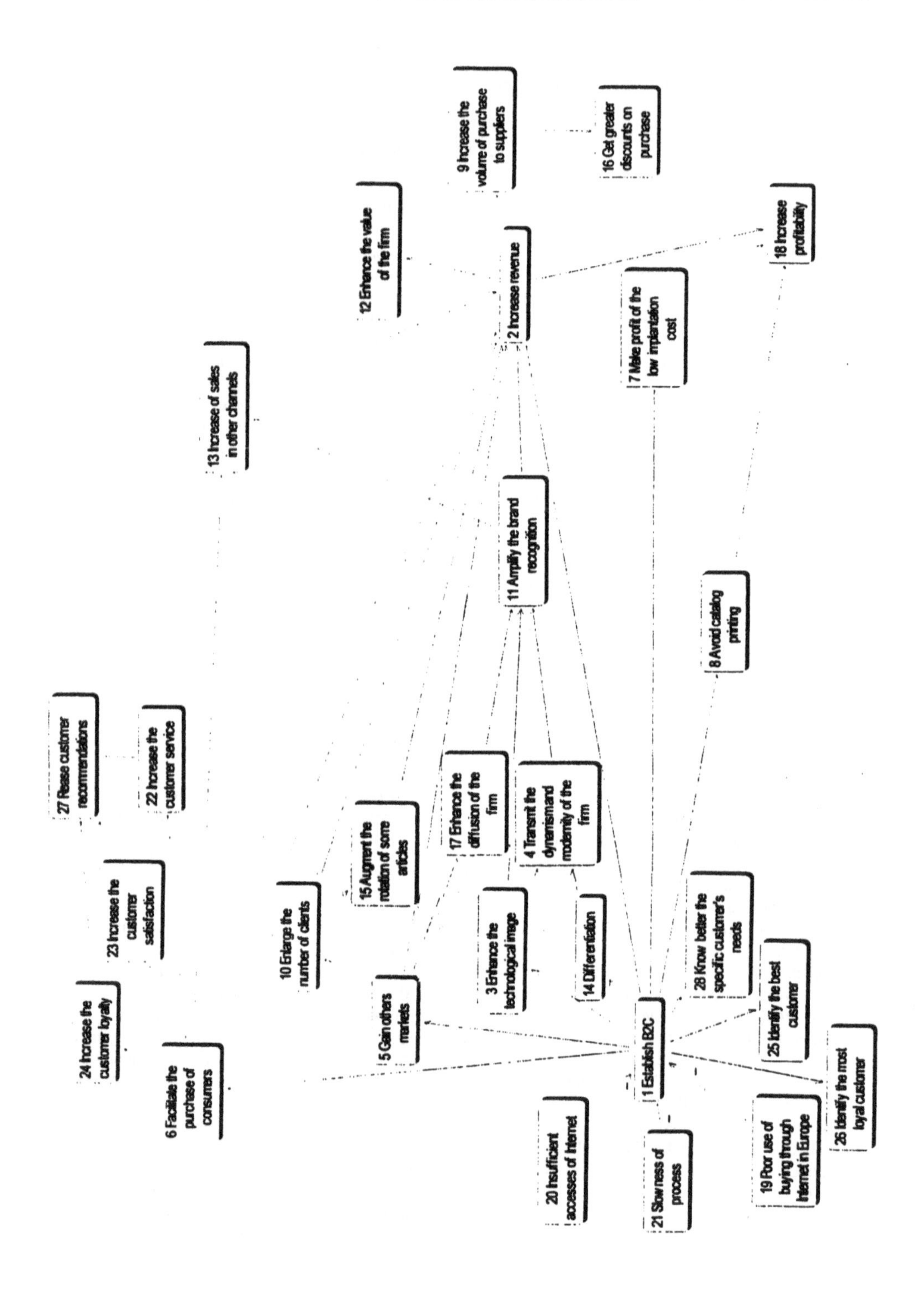

Figure 1: Cognitive Map for B2C systems adoption.

5. Conclusion

This research reveals how the knowledge of the top manager cognitive process can help in the decision making process. It offers the means to obtain the true expectations behind the B2C implementation. If we know the real requirements, getting project success will be easier for developers.
In relationship with studies made about the reasons to establish B2C in other industries, our study offers different results because we use methods that can extract the true motivations that are in the decision makers' minds and the relative importance that every construct has comparatively with the others.
B2C can add significant competitive advantages with cost reductions. However, in our study the creation of most important competitive advantages comes from growth and differentiation.
The main contribution of this research is identifying the true reasons that decision makers have to establish B2C. It can improve the project team's knowledge about requirements. It is a critical issue in any technological project. The findings can offer a number of important implications for practitioners and academics working within the B2C.
The situation facing this study can be used as a frame to apply to other IT projects in the fashion industry and analogous industries which would benefit from this paper.

Acknowledgments

The author wishes to thank to Sebastian Lozano for his fruitful comments and suggestions.

References

[1] R. Calori, G. Johnson and P. Sarnin, "CEOs' Cognitive Maps and The scope of the organization". *Strategic Management Journal* 15(1994), pp. 437-457.

[2] K. Changa, J. Jackson and Grover V. "E-commerce and corporate strategy: an executive perspective". *Information & Management* 40(2003), pp. 663-675.

[3] D. Chatterjee, R. Grewal, and V. Sambamurthy "Shaping up for e-commerce: institutional enablers of the organizational assimilation of web technologies". MIS *Quarterly* 26(2002), pp. 65-89.

[4] C. Eden, and F. Ackermann, "Strategy development and implementation – the role of a group decision support system" in Holtham, C. *Executive Information Systems and Strategic Decision Support*, Unicom, Uxbridge, UK, (1992) pp. 53-77.

[5] G. Kelly, "*A brief introduction to personal construct theory*". Academic Press, London (1970).

[6] G. Kelly, "*The Psychology of Personal Constructs*". New York, Norton & Company (1955).

[7] M.R. Subramani, and E. Walden "*The Impact of e-commerce Announcements on the Market Value of Firms*", Information Systems Research 12(2001), pp. 135-154.

Brill Academic Publishers
P.O. Box 9000, 2300 PA Leiden,
The Netherlands

Lecture Series on Computer and Computational Sciences
Volume 2, 2005, pp. 120-123

Detecting IS/IT Future Trends: An Analysis of Technology Foresight Processes Around the World

J. L. Salmeron and V. A. Banuls[1]

University of Pablo de Olavide,
Crtra. de Utrera, km. 1, Seville (Spain)

Received 8 April, 2005; accepted in revised form 21 April, 2005

Abstract: The research presented here is meant to (1) place relevance on the national IS/IT research priorities (2) bring researchers and practitioners a global reference about IS/IT future trends and developments. To accomplish the goal, a content analysis of 15 national foresight processes is done. Through this analysis, the authors recognize different IS/IT concerns in each county. Nevertheless, some IS/IT developments are seen as critical at the world level: ubiquitous computing, user interfaces and broadband-mobile networks.

Keywords: Technology Foresight, Information Systems, Information Technology, Content Analysis.

1. Introduction

The detection of technological trends in the Information Systems and Information Technology (IS/IT) industry imposes a constant challenge for academics, practitioners and policy makers. On the one hand the IS/IT industry is a key driver of economic growth. On the other hand is the dynamism of the IS/IT industry.

In this critical and dynamic context, Technology Foresight (TF) arises as an instrument of strategic policy intelligence in order to manage the uncertainty of the IS/IT environment [14]. In general, TF processes orientate towards the longer-term future of science, technology, economy and society, with the aim of identifying the areas of strategic research and the emerging generic technologies likely to yield the greatest economic and social benefits [10].

In TF processes, IS/IT has been treated as a central key area [5,13]. In spite of the main role of IS/IT in TF, there are no links between the IS/IT literature and TF processes. In this work, the authors aims to contribute to link both fields with an analysis of the IS/IT role in TF. This contribution is something new in the literature.

This paper is structured as follows. Section two analyses the IS/IT foresight background. The third section seeks to analyze the research design. This paper concludes with the discussion of results and conclusions.

2. IS/IT Foresight Background

IS/IT has often been treated in the TF process from a national viewpoint. The first foresight initiative that include IS/IT as a key area was the Japanese foresight program [12]. In this programme the Japanese National Institute of Science and Technology Policy (NISTEP) represents the future Japanese society by means of the technological trends, being IS/IT a key area from its beginning in 70's. Along those lines, almost all countries with a foresight program, include IS/IT as a key area (i.e. France [4], United Kingdom [9], Australia [2] and U.S.A. [11] to name a few). As a result of these processes, IS/IT trend lists form and, scenario and/or recommendations are presented. To reach their goal, TF processes involve experts from technological, economical and policy making sectors as well as other scientific fields.

Although there are many foresight processes that highlight the relevance of IS/IT in society and the economy, there are no links between foresight and IS/IT literature. This lack of relationships between both fields implies that (1) IS/IT foresight is not fundamental in IS/IT literature (2) IS/IT foresight results are for hidden for IS/IT researchers.

[1] Corresponding author: vabansil@upo.es (V. A. Banuls).

In this work we want to contribute to advance this relationship between both fields, through the analysis of IS/IT foresight results based on categories coming from IS/IT literature. In next section the research design completed to reach the goal is analysed.

3. Research Design

The research design is based on a content analysis of the IS/IT foresight findings. This methodology is selected in order to manage the heterogeneity of IS/IT foresight reports output. In this section the main issues of the application are the methodology to study, which are categorization, sample and analysis, are shown.

A. Categorization.

The categorization on which the work is based comes from an IS/IT literature review. Specifically, we are going to analyze the IS/IT foresight reports taking as reference the validated key areas of Straub and Wetherbe's work about the future of Information Technology on 90's [16]. In this paper, the key areas were "Human Interface Technologies", "Communication Technologies" and "System Support Technologies". The reasons for selecting this paper are because (1) it is the most referenced in IS/IT trends literature (2) this paper was validated by Grover and Goslar [8] for IS/IT future research.

B. Sample.

A sample of the most important national TF activities was analyzed. The initial size sample of TF exercises was 74, and finally was reduced to 15. This reduction was done in order to take the last national exercises, with IS/IT with a key area and with results that allow its processing.

C. Analysis.

The classification of each IS/IT into categories was validated in discussion and co-coding comparisons between the two authors and refined during several passes through the discourse. Intercoder reliability was calculated by determining the number of agreements and disagreements in the application of the codes to the transcripts. Overall intercoder reliability was over 85%, being a satisfactory percentage.

4. Discussion

Through the content analysis of over 300 critical technologies in 15 national foresight exercises, the authors recognize different IS/IT concerns in each county (Table 1). First there are countries that are focused on Support Systems Technologies [Holland (software engineering), France (components) and USA (components, software engineering and computer systems)]. Other countries are mainly focused on communication technologies [India (communications base technologies) and Japan (communications applications)]. The rest of the countries are maintaining equilibrium between these categories and the Human Computer and Interaction Technologies. These differences are mainly due that the critical technology focus, depends on the industry and/or societal needs relevant for each country [5,7].

Table 1: IS/IT Trend Map

Country	Program	Human Interface Technologies	Communication Technologies	System Support Technologies
South Africa	National Research and Technology Foresight Project	29%	14%	57%
USA	New Forces at Work: Industries Views Critical Technologies	9%	13%	78%
India	Technology Vision 2020	0%	69%	31%
Japan	7th Technology Foresight	8%	77%	15%
South Korea	Second Technology Forecast	0%	52%	48%
Czech Republic	Analyses of international key technologies lists	0%	17%	83%
Spain	Industrial Technology Foresight reports	0%	50%	50%
Finland	On the way to technology vision	14%	29%	57%
France	List of Key Technologies	3%	17%	80%
Ireland	Technology Foresight Ireland	22%	39%	39%
Israel	Science and Technology Foresight for Israel	29%	43%	29%
Sweden	Second Teknisk Framsyn	14%	57%	29%
Holland	Technology Radar	33%	0%	67%
UK	The Foresight Program	20%	26%	54%
Australia	Matching science and technology to future needs 2010	0%	67%	33%

A second main result of the study is the critical IS/IT list per each category (Table 2). In this critical list are ranked the five most relevant IS/IT in each category. This classification has been done by the exhaustive analysis of the TF reports.

Table 2: IS/IT Critical List

Human Interface Technologies	1. User Interfaces. 2. Multimedia working. 3. Virtual reality. 4. Speech and image recognition. 5. Improved displays for human computer interaction.
Communication Technologies	1. Broadband networks. 2. Mobile technologies. 3. Wireless communications. 4. Improved technologies for the security of networks. 5. Optical transmission systems.
System Support Technologies	1. Ubiquitous Computing 2. Components (Sensors and Semiconductors). 3. Software engineering tools. 4. Data and knowledge processing. 5. Storage technologies.

Following this classification, the most important developments in the IS/IT field will be in the ubiquitous computing, user interfaces and broadband-mobile networks. Moreover, it is important to remark that in expert's opinion that advances will be done in the time horizon 2015. This asseveration can be observed in conclusion of several TF reports [12,15]. Once this period came, scientific priorities in the IS/IT will change to specifics applications to life styles, materials and biotechnology.

5. Conclusions

In this work, the authors aims to contribute to the IS/IT literature with an analysis of the IS/IT role in TF. With this study the authors aim to show TF results to IS/IT community. Specifically, those results throw that in next decade IS/IT will be a mature sector. Meanwhile, national R&D efforts have to be focused on making systems more usable and connected. In this task, each country will be still focused in a part of it. Mainly ones will be focused on the systems support, others on communications developments.

Once pass the time horizon 2015, in expert's opinion IS/IT converge with other sciences. This tendency can also be observed if we examine the growing interactions between biotechnology, nanotechnology and IS/IT fields. For that reason national research networks need to develop the appropriate scientific capacity for connecting R&D inputs.

References

[1] Applewhite, A., Kumagai, J., Technology Trends 2004. *IEEE Spectrum*. January 14-19 (2004).

[2] ASTEC, Developing Long Term Strategies for Science and Technology in Australia: Outcomes of the Study-Matching Science and Technology to Future Needs 2010, Australian Goverment Printing Service, Camberra (1996).

[3] Benjamin, R., Information technology in the 1990s: A long-range planning scenario. *MIS Quarterly* 6 (2) 11- 31 (1982).

[4] CM International, Technologies Clés 2005, Ministère de l'Economie, des Finances et de l 'Industrie, Secretariat d'Etat à l'Industrie, Service de l'Innovation 99 (2000) *(in French)*.

[5] Ducatel, K., Information and communication technologies: future perspectives on ICTs *Foresight*, 1 (6) 519-535 (1999)

[6] Foren, A practical Guide to Regional Forecasting. European Commision Research Directorate General, European Communities (2001).

[7] Gavigan, J.P., Scapolo, F., Matching methods to the mission: a Comparison of national foresight exercises, *Foresight* 1 (6) 495-517 (1999).

[8] Grover, V, Goslar, M., Information technologies for the 1990s: the executives' view, *Communications of the ACM 36* (3) 17-19 (1993).

[9] Keenan, M., Identifying emerging generic technologies at the national level: the UK experience, *Journal of Forecasting* 22 (2-3) 129-160 (2003).

[10] Martin, B., Foresight in science and technology, *Technology Analysis and Strategic Management* 7 (2) 139 – 168 (1995).

[11] National Critical Technologies Panel, National Critical Technologies Panel Report, U.S. Government, Washington D.C. (1998).

[12] NISTEP, The Seventh Technology Forecast Survey - Future Technology in Japan, NISTEP Report No. 71, Tokyo (2001).

[13] OCDE, Special Issue on Government Technology Foresight Exercises. STI Review 17 (1996).

[14] Salo, A. Cuhls, K., Technology Foresight – Past and Future, *Journal of Forecasting* 22 (2-3) 79 – 82 (2003).

[15] Shin, T., Using Delphi for a Long -Range Technology Forecasting, and Assessing Directions of Future R&D Activities. The Korean Exercise, *Technological Forecasting and Social* Change 58 (1-2), 125-154 (1998).

[16] Straub D.W, Wetherbe, J.C., Information technologies for the 1990s: an organizational impact perspective, *Communications of the ACM* 32 (11) 1328 - 1339 (1989).

[17] Wagner, C.S., Popper, S.W., Identifying Critical Technologies in the United States: a Review of the Federal Effort, *Journal of Forecasting* 22 (2-3) 113–128 (2003).

Brill Academic Publishers
P.O. Box 9000, 2300 PA Leiden,
The Netherlands

Lecture Series on Computer and Computational Sciences
Volume 2, 2005, pp. 124-127

Modelling Success in Information Technology Projects

Jose L. Sameron and M. Baldomero Ramirez[1]

University Pablo de Olavide
41013 Seville, SPAIN

Received 11 April, 2005; accepted in revised form 21 April, 2005

Abstract: The use of new methodologies is needed for the success in the Information Technology projects in organizations. This research designs a tool to support this objective. This model is presented as an innovative proposal to reduce the limits of the success modelling. This work contributes to the formalization and objectivity of the success models in IT projects. Overall the aim is to advance in an almost unexplored direction.

Keywords: Success, Modelling, IT projects, MCDM, e-Learning.

1. INTRODUCTION

The use of new methodology is needed for the success in the Information Technology (IT) projects implementation. This research designs a tool to support this objective. This model is presented as an innovative proposal to reduce the limits of the success modelling.

This work is based on the identification, classification and sequence of the attributes and objectives of the IT projects through the application of a multicriteria tool.

2. MODELLING SUCCESS

The traditional Critical Success Factors (CSFs) approach relates the company's objectives assuring the success of its proposed goals (Rockart, 1979). Salmeron and Herrero (2005) suggest that the CSFs can be modelled in hierarchical structures through the application of formalized quantitative methodologies. Therefore, an additional value is CSFs added to the traditional. The Critical Successful Chains (CSCs) are included in this concept because they include the former concepts and make reference to the link between Information System attributes, CSFs and the organization's goals (Peffers and Gengler, 2003). These chains model CSFs as graphs not as hierarchical structure.

To obtain clearer information about the project that would contribute to its successful introduction implementation the implicit relationship between the factors should be made explicit through the collaboration of the members of the projects (Earl, 1993, Gottschalk, 1999 and Hackney, *et al.*, 1999). In fact, it is possible to model chains referred to different points of view within the project. Obviously, we would not obtain the same successful model based on the user than realized with the designers. An integrative approach to this methodology requires the modelling of the chains under every possible perspective. However, this work realizes the modelling of the CSC related to the designer.

In addition, an improvement of the original CSCs method is applied centred in the information elicitation phase. The process is held without the use of interpretative procedures of experienced analysts. Concretely, we have selected a multicriteria methodology analysis of content, due to the fact that it realizes an objective description, systematic and quantitative of the content make explicit in the communication.

[1] Corresponding author. e-mails: jlsalsil@upo.es, mbramfer@upo.es

3. CASE STUDY

In this case study, we are going to analyze the e-Learning designer success model using improved CSCs. The authors have selected this kind of project because we consider that it is an emerging and interesting field within IT area.

The questionnaires where addressed to the designer, instructors and students from an e-learning project. It has to be taken into account that every project participant whose CSCs is being modelled must answer the questionnaires from every point of view. The inclusion of instructors and students is relevant because they offer an external view different than designer's view.

With the results valid inferences are hold, to be applied in their own context through the analysts of content technique. This research methodology is the most adequate to the recompilation of information, to the extent that it makes explicit a person, group or community interests, outside its context (Krippendorff, 1990).

The process of the analysis of content was developed in the universe definition, establishment and definition of the units of analysis, the categories and subcategories that present the research variables, coders, elaboration of the codification sheets and the conclusions of the codification. The criteria for the evaluation of the methodology quality are based in the validity and reliability of the analysis.

Reliability will be measured with coefficient C.R.

$$C.R = \frac{2m}{N_1 + N_2} \quad (1)$$

where:

"m" is the number of codification decision where coder 1 and coder 2 coincide.
"N_1" is the number of codification decision of coder 1.
"N_2" is the number of codification decision of coder 2.

The reliability coefficients of the analysis of content calculated in (1) for each item surpassed el 0.8. It implies a high level of reliability in the codification. In this work it only makes sense to talk about a validity oriented to the data, because it evaluates to what extent a method of analysis is representative of the information inherent to the available data.

Once all the questionnaires where codified, the authors agreed in a list of 28 elements integrating the CSC for the designer.

In the model each elements is interpreted as a node and a value inside it that represents the number of participants in the questionnaire that name mentioned it. The circle area is proportional to the value that contains.

The analysts have interpreted the possible causal sequences between the different nodes of de CSC and the edges between the graph nodes where represented. These edges offer an added value to the enumeration of the nodes, because it allows the identification of its interrelationships.

The chain models the sequence of the explicit information in a graph. Hence, the success model is formalized from the e-learning project designer's perspective.

4. DISCUSSION AND CONCLUSIONS

The graph must be interpreted like a sequence of the codified ideas that it develops the CSCs in the e-Learning designer activities.

This model begins in the nodes located in the zone of the left of the graph and finishes in those of the right of the same one.

According to the interpretations of the participants, designer needs like fundamental attribute that the project is hot topic. In order to obtain this node, one is due to use suitable tools of communication, a center with guarantees, a course that has an accreditation and a suitable promotion of the same one.

The primary object of the e-Learning project is based on a competitive knowledge level of the students. The structuring of the contents is the CSF that must have an e-Learning project so that the students reach a competitive knowledge level and is hot topic. This factor complements with other CSFs like the facility of access to the project and the designer compensation within expectations of the course.

From structured contents affluent, the sequence of CSFs that reach a competitive knowledge level can be made of two different ways. On the one hand, contents based on practical cases and by the other, a significant learning will be used that will be needed it develops a rate of increasing learning in difficulty.

The innovation in the education process is an objective of the project originated by the competitive knowledge level of the students. As well, this innovation develops to a sequence of objectives for the designer like the enrolment of prospective students and the quality of education.

This way, the chain of left to right is crossed sequentially until the attainment of the objectives of the e-Learning project and the successful implementation. The graphical vision of the chain detects the relations between nodes immediately and fortifies the project strategically considering the greatest nodes.

This work develops a tool based in the content analysis methodology within the CSC approach, reducing the analyst's interpretations and disentangling the information from the project environment.

This work contributes to the formalization and objectivity of the success models in IT projects. Overall, the aim is to advance in an almost unexplored direction. The authors are working on future research projects about the formalization of the different phases of the CSCs design process.

References

[1] Earl, M.J. *Experiencies in strategic information systems planning.* MIS Quarterly 7(1), pp. 1-24. 1993.

[2] Gottschalk, P. *Strategic information systems planning: the IT strategy implementation matrix.* European Journal of Information Systems 8, pp. 107-118. 1999.

[3] Hackney, R., Kawalek, J. y Dhillon, G. *Strategic information systems planning: perspectives on the role of "end-user" revisited.* Journal Of End User Computing 11(2), pp. 3-12. 1999.

[4] Krippendorff, K. *Content analysis. An Introduction to its Methodology.* Beverly Hills: Sage. 1980.

[5] Peffers, K. and Gengler, C.E. *How to identify new high-payoff information systems for the organization.* Communications of the ACM. 46 (1), pp. 83-89. (2003).

[6] Rockart, J.F. *Chief executives define their own data needs.* Harvard Business Review 52(2), pp. 81-93. (1979).

[7] Salmeron, J.L. y Herrero, I. *An AHP based methodology to rank critical success factors of executive information systems.* Computers Standards & Interfaces. Forthcoming. 2005.

APPENDIX

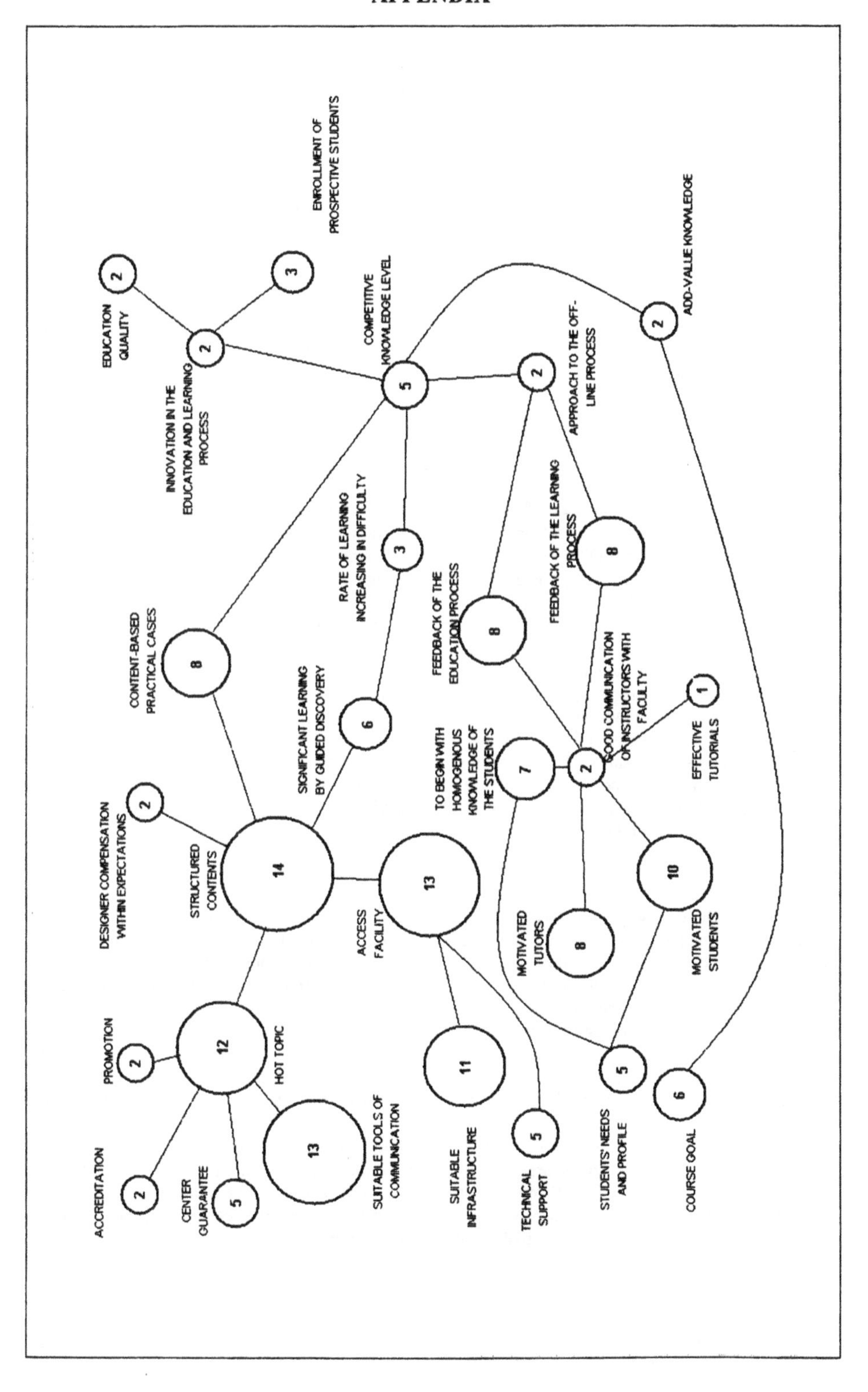
ACCREDITATION
CENTER GUARANTEE
PROMOTION
HOT TOPIC
SUITABLE TOOLS OF COMMUNICATION
DESIGNER COMPENSATION WITHIN EXPECTATIONS
STRUCTURED CONTENTS
ACCESS FACILITY
SUITABLE INFRASTRUCTURE
TECHNICAL SUPPORT
CONTENT-BASED PRACTICAL CASES
SIGNIFICANT LEARNING BY GUIDED DISCOVERY
RATE OF LEARNING INCREASING IN DIFFICULTY
EDUCATION QUALITY
INNOVATION IN THE EDUCATION AND LEARNING PROCESS
ENROLLMENT OF PROSPECTIVE STUDENTS
COMPETITIVE KNOWLEDGE LEVEL
APPROACH TO THE OFF-LINE PROCESS
ADD-VALUE KNOWLEDGE
FEEDBACK OF THE EDUCATION PROCESS
FEEDBACK OF THE LEARNING PROCESS
TO BEGIN WITH HOMOGENOUS KNOWLEDGE OF THE STUDENTS
GOOD COMMUNICATION OF INSTRUCTORS WITH FACULTY
EFFECTIVE TUTORIALS
MOTIVATED TUTORS
MOTIVATED STUDENTS
STUDENTS' NEEDS AND PROFILE
COURSE GOAL

Brill Academic Publishers
P.O. Box 9000, 2300 PA Leiden,
The Netherlands

Lecture Series on Computer and Computational Sciences
Volume 2, 2005, pp. 128-131

An Isomorphic-Based Model in IT/IS Assimilation

J.L. Salmeron and Salvador Bueno[1]

University Pablo de Olavide,
Carretera de Utrera, km. 1, Seville (Spain)

Received 8 February, 2005; accepted in revised form 16 March, 2005

Abstract: The isomorphic view offers an approach to understanding IT/IS diffusion and the adoption process caused by isomorphism within a specific environment, mainly industry. This proccess can be an answer to the pressures that organizations receive to be similar. With the purpose of analyzing this proccess, we have, using multi-variant analysis, identified an IT/IS adoption typology through a sample of organizations coming from main industries. This typology has allowed us to evaluate IT/IS isomorphism, and to analyze the explanatory potential of the isomorphic approach in order to evaluate IT/IS assimilation in organizations.

Keywords: IT/IS industry-based classification, isomorphism, industry, multivariate

1. Introduction

The role attributed to Information Technologies and Information Systems (IT/IS) by organizations has evolved over the last decades from a passive direction to a strategic and competitive one [1]. Literature confirms [14] that IT/IS help organizations in business-strategy implementation as well as in reaching their goals. Furthermore, it is extracted that fitting an organization with an IT/IS increases its competitive advantage [2], [4], [5], [12].

Moreover, in many situations IT/IS are considered a crucial element in industries in order to compete [10], [13]. In this way, the isomorphic view considers environmental characteristics to be one of the main factors that compel IT/IS implementation in organizations. For this reason, the present study takes on an isomorphic approach. Based on this, it is supposed that organizations in the same industry eventually reach the same level of IT/IS assimilation.

Our investigation takes on this idea and our objective is to determine if, in fact, the isomorphic view can explain the assimilation of IT/IS in organizations. The research content, including an analysis of the most relevant points concerning this view, as well as its relationship to IT/IS adoption, are presented in section 2. In section 3, the research methodology is also described. Lastly, conclusions and future research are suggested.

2. Isomorphic process

The isomorphic view point offers a framework to understanding IT/IS diffusion and the adoption processes. This approach affirms that structure and activity modifications are not always made under purely economic criteria, but that they sometimes respond to pressures coming from the environment.

One of the central aspects of the isomorphic view is the imitative process with respect to the different enviromental agents. This imitation is based on the probability of organizational survival, mainly in industries with a high uncertainty. This process is known as Isomorphism [8], [15].

Isomorphism within the institutional environment has some crucial consequences for the organization. Therefore organizations tend to incorporate elements that allow them to legitimize themselves externally, forgetting their efficiency. This however, does not necessarily mean that they are inefficient [6].

This isomorphic process tries to explain the similarity among organizations assigned to a certain scope of industry. It puts a lot of pressures on organizations and it is very probable that companies found within the same industry will adopt similar IT/IS. The industry-based distinction is motivated by the relevance that the isomorphic approach grants to industry [7]. So, it is considered that companies operating in a same environment tend to develop similar isomorphic activities.

A plausible connection between IT/IS adoption/use and the institutional theory is identified in the literature. The goal of this research is to study this relationship.

[1] Corresponding author.
E-mail: sbueavi@upo.es. Phone +34 954349063. Fax: +34 954348353

3. Research methodology

The objective of this empirical research is to identify an IT/IS adoption typology to answer the hypotheses of this study. The sample used is from 163 organizations, corresponding to 6 divisions: Agriculture and Food (6), Manufacturing (50), Construction (13), Transportation (8), Distribution (53) and Services (33). The selection sample is justified by the industrial representation of the analyzed organizations within their geographical region.

A questionnaire was the tool used to obtain the information. It was designed for the directors of the IT/IS departments.

4. Industry-based evaluation of IT/IS assimilation

In this section, the evaluation of IT/IS assimilation in industries is analyzed. To reach this objective, a multivariate analysis has been used. In order to apply this analysis, the most relevant variables in each one of the IT/IS aspects contemplated in the questionnaire have been selected. After making several tests, the variables that best explain IT/IS use are those that identify the presence of each one of the IT/IS. Specifically, the variables are (1) existence of IT/IS departments [x_1], (2) existence of ERP [x_2], (3) existence of Internet [x_3], (4) existence of websites [x_4], (5) sales through Internet [x_5] and (6) purchasing through Internet [x_6].

The variables are binary. That is to say, they can take only two values, yes or no. With the purpose of incorporating these variables into the cluster analysis, a numerical value has been assigned to each one of the two positions (one meaning "yes", and zero meaning "no").

A. Previous analysis.

The hierarchical cluster analysis is applied to determine the suitable number of clusters that will finally be considered. The results identify four clusters which are discussed in the results as being most suitable. Then, ANOVA analysis is applied in order to observe the presence of any significant differences among clusters.

This analysis contrasts the differences of the mean values that the variables take [3]. With this objective, the F statistical is used at a critical level of 0.05 ($p<0.05$) to verify that all the variables included in the analysis take on different values for each one of the groups formed. The analysis confirms that all the variables are significant for $p<0,05$, and thus, confirming its suitability.

Furthermore, the discriminant analysis shows that all the values are found to be significant at a critical level of 0.05, which indicate that the four clusters are sufficient to explain IT/IS use. Additionally, this analysis shows that 94.5% of the cases have been classified correctly and allows us to affirm the high explanatory power of the model.

B. Cluster composition.

The objective of the k-means cluster analysis is to determine which companies, and mainly which industries, will comprise each one of the clusters (Table 1). These values will help to determine IT/IS assimilation followed in each cluster.

Table 1: Final values.

Variable	Description	Cluster			
		1	2	3	4
x_1	Existence of IS/IT departments	1	1	0	0
x_2	Existence of ERP	1	0	0	0
x_3	Existence of Internet	1	1	0	1
x_4	Existence of websites	1	0	0	1
x_5	Sales through Internet	0	0	0	0
x_6	Purchasing through Internet	0	0	0	0

5. Results

From the results obtained, four IT/IS assimilations have been identified and to which each one, a term has been assigned. This terminology is similar to the strategic typology proposed by [11]. For the allocation of these terms, the importance of having an IT/IS department has been highlighed [9]. Below the four different clusters are identified.

IT/IS assimilation of the first cluster is innovative, in that it has implanted most of the IT/IS considered in the analysis. This cluster makes up of 22.70% of the cases.

On the other hand, defensive assimilation of IT/IS has been assigned to the second cluster, in spite of having a computer science departments. These companies have only implanted the most stable and permanent IT/IS in business, Internet being a good example. This second cluster represents the most numerous group, made up of 46.62% of the cases analyzed.

The third cluster takes on passive IT/IS assimilation. These companies do not consider IT/IS relevant for their activity. They do not implement any of the analyzed IT/IS in this study. This group is a minority, identifying with only 9.81% of the companies.

Finally, the fourth cluster has an IT/IS assimilation following in spite of not having an IT/IS department. These companies have, however, adopted technologies like Internet and websites. The use of websites indicates that these companies, and those of cluster one, are begining to consider IT/IS use as a way to develop their business activities. This group comprises 20.85% of the cases.

As an industrial classification the composition of each cluster is very heterogenous and it is therefore quite difficult to identify IT/IS adoption tendencies by industries (Table 2).

Table 2: Clusters in each industry

Industry	Agriculture and Food				Construction			
N	6				13			
Cluster	1	2	3	4	1	2	3	4
N	0	4	1	1	1	11	0	1
%	0.00%	66.67%	16.67%	16.67%	7.69%	84.62%	0.00%	7.69%
Industry	Distribution				Manufacturing			
N	53				50			
Cluster	1	2	3	4	1	2	3	4
N	14	23	6	10	14	22	5	9
%	26.42%	43.40%	11.32%	18.87%	28.00%	44.00%	10.00%	18.00%
Industry	Services				Transportation			
N	33				8			
Cluster	1	2	3	4	1	2	3	4
N	8	14	4	7	0	2	0	6
%	24.24%	42.42%	12.12%	21.21%	0.00%	25.00%	0.00%	75.00%

The results reached only partially reflect the results expected. Moreover, they do not demonstrate that the majority of the organizations adopt the same IT/IS use. Nevertheless, it is possible to identify similarities with respect to the identified adoptions in the study.

On the basis of the results obtained, conclusions from the agriculture and food, construction, and transportation industries must be taken with caution, due to the fact that only a reduced volume of organizations has been analyzed. For that reason, it would be advisable to extend the study in future research of these industries in order to verify if the same proportion is maintained when increasing the sample size.

6. Conclusion

The results of this investigation demonstrate the existence of strong pressures within a sector that tend toward adopting a specific IT/IS. Based on the results, we have observed how the environment, in each one of the industries, is capable of putting pressure on organizations so that specific IT/IS are implemented. This is a consequence of the isomorphic activity development within the sector. We consider that the incorporation of isomorphism provides a relevant contribution to the studies related to IT/IS.

Furthermore, we have reached significant conclusions regarding the relationship that exists between environmental characteristics and IT/IS assimilation. Basically, we have observed the existence of the tendency to implement the same IT/IS throughout companies that belong to the same industry.

We consider that it is in this last aspect where a greater contribution to the related research is offered with respect to the isomorphic view. This is due to the fact that the isomorphic process application in the field of IT/IS has been quite limited. In conclusion, we consider that the isomorphic view offers an explanatory potential to understanding the adoption and use of IT/IS in organizations.

References

[1] Avison, D.E., Cuthbertson, C.H. and Powell, P. *The paradox of information systems: strategic value and low status.* The Journal of Strategic Information Systems 8 (4) 419-445 (1999).

[2] Byrd, T.A. and Turner, D.E. *An exploratory examination of relationship between flexible IT infrastructure and competitive advantage.* Information & Management 39 (1) 41-52 (2001).

[3] Cortina, J.M. and Nouri, H. *Effect size for ANOVA designs.* Sage, London, 2000.

[4] Davis, L., Dehning, B. and Stratopoluos, T. *Does the market recognize IT-enabled competitive advantage?* Information & Management 2020:1-12 (2002).

[5] Dehning, B. and Strapoulos, T. *Determinant of a sustainable competitive advantage due to an IT-enabled strategy.* Journal of Strategic Information Systems 12 (1) 1-22 (2003).

[6] DiMaggio P. J. and Powell, W.W. *The iron cage revisited: institutional isomorphism and collective rationality in organizational fields.* (Editors: W. Powell & P. DiMaggio), The new institutionalism in organizational analysis (1991), 63-82. The University of Chicago Press, Chicago.

[7] DiMaggio, P. *Structural analysis of organizational fields: a Blockmodel approach.* Research in Organizational Behaviour 8: 335-370 (1986).

[8] DiMaggio P.J. and Powell, W.W. *The iron cage revisited: institutional isomorphism and collective rationality in organizational fields.* American Sociological Review 48 (2) 147-160 1983.

[9] Jiang, J.J., Klein, G. and Pick, R.A. *The impact of IS department organizational environments upon project team performances.* Information & Management 40 (3) 213-220 (2003).

[10] Larsen, T.J. and Wetherbe, J.C. *An exploratory field study of differences in information technology use between more- and less-innovative middle managers.* Information & Management 36 (2) 93-108 (1999).

[11] Miles, R.E. and Snow, C. *Designing Strategic Human Resources Systems.* Organizational Dynamics 5: 36-52 (1984).

[12] Mykytyn, K., Mykytyn, P.P., Bordoloi, B., McKinney, V. and Bandyopadhyay, K. *The role of patents in sustaining IT-enabled competitive advantage: a call for research.* Journal of Strategic Information Systems 11 (1): 59-82. (2002).

[13] Raymond, L. *Determinants of Web site implementation in small business.* Internet Research: Electronic Networking Applications and Policy 11 (5): 411-422 (2001).

[14] Salmela, H. and Spil, T. *Dynamic and emergent information systems strategy formulation and implementation.* International Journal of Information Management 22 (6): 441-460 (2002).

[15] St. John, C.H., Cannon, A.R. and Pouder, R.W. *Change drivers in the new millennium: implications for manufacturing strategy research.* Journal of Operations Management 19 (2): 143-160 (2001).

Brill Academic Publishers
P.O. Box 9000, 2300 PA Leiden,
The Netherlands

Lecture Series on Computer and Computational Sciences
Volume 2, 2005, pp. 132-135

Artificial Intelligence Hybrid Model Applied to Wastewater Treatment

Yanet Rodríguez Sarabia[1], Xiomara Cabrera Bermúdez[2], Rafael Jesús Falcón Martínez[1], Zenaida Herrera Rodríguez[2], Ana M. Contreras Moya[2], Maria Matilde García Lorenzo[1]

Central University of Las Villas
Santa Clara, Cuba

Received 7 April, 2005; accepted in revised form 22 April, 2005

Abstract: This paper refers to a way of proposing the optimal sequence of treatments that should be applied to wastewater by using a hybrid model. It combines cases-based reasoning and artificial neural networks, getting the best of both approaches. Preliminary results demonstrate that it is a feasible model.

Keywords: wastewater treatment, neural network, cases-based reasoning, hybrid model.

1. Wastewater treatment

The hydric resources of our country have been affected for the industrial and domestic wastewater emissions, some of them with deficient treatments. Wastewaters have contaminant properties that can be transmitted to the final receptor. In order to know the contaminant power of wastewaters it is necessary to study theirs physical, chemical and biological properties. Besides, for designing, control and good operation of the wastewater treatment plants is important to determine of such parameters.

Among the common parameters [1], for an adequate characterization we can find the followings:

- Temperature: It is important due to its effect on the aquatic life, on the chemical reactions and on the development of microorganisms with the perturbation in the treatment process.
- pH: This is the measure of the water's acidity once it leaves the plant. Ideally, the water's pH would match the pH of the river or lake that receives the output of the plants.
- BOD5 (biological oxygen demand): It is a measure of the strength of the wastewater. BOD5 is a measure of how much oxygen in the water will be required to finish digesting the organic material left in the effluent.
- COD (Chemical Oxygen Demand): This is the amount of oxygen necessary to oxidize the organic matter present in the water using a strong oxidant agent.
- Solid content: Its determination is very useful to determine the best technological process of depuration.
- Total nitrogen: This is the measure of the nutrients remaining in the water. The nitrogen compounds have a great interest for their influence in the life processes of all plants and animals. The nitrogen contents are extremely important in liquid waste treatments because there must be a relation with carbon and phosphorous compounds content in order to obtain an efficient biological treatment.
- Total phosphorous: This is the measure of the nutrients remaining in the water. Phosphorous is essential for the growth of algae in lagoons and rivers but if this element exists in a great quantity jointly with Nitrogen, the Eutrophication can be produced.

[1] Computer Science Department Emails: yrsarabia@uclv.edu.cu, rfalcon@uclv.edu.cu, mmgarcia@uclv.edu.cu
[2] Environmental Virtual Institute. Emails: qfx@uclv.edu.cu, zenah@uclv.edu.cu, anama@uclv.edu.cu

Many efforts have been done for diminishing this contamination. Before making any external treatment to the wastes it is necessary to take into account where these treated wastes are going to be spilled. The norms of spill and an analysis of the general process will allow diminishing wastes streams and reusing these in different parts of itself industry.

In the case where the necessity of extern treatment is imminent, the analysis of waste characteristic is the starting point. It is important to take in account the contaminant charge magnitude, the pH of the waste stream, the BOD5/COD relation, the quantity of solids and its quality, it means to know if it is possible that they can settle or not.

It is necessary to indicate that it is very important to take into consideration the criteria of experts for deciding the optimal sequence of treatments. In spite of this, some treatment sequences for the study object industries are exposed. It is clarified, that before to the proposed treatments for each type of industry, it is necessary to include the preliminary screening to remove large suspended solids, metal and rags because this kind of materials could obstruct the following stages of treatment. Normally it is based on physical procedures that include the sifting, equalization and sand elimination.

2. Feasibility of applied Artificial Intelligence techniques

In the past few years, artificial neural networks (ANN) have been used in describing and modeling wastewater treatment processes. An exhaustive review of the usage of ANN models in environmental problems was made in [2][3]. For example in 1991, Capodaglio applied a feed-forward back propagation network to the analysis of bulking conditions at a wastewater treatment plant (WWTP); in 1994 an ANN was used by Hung to predict the performance of wastewater treatment plant.

Case-based reasoning (CBR) systems have been used in a broad range of domains to capture and organize past experience and to learn how to solve new situations from previous solutions. CBR systems have been applied for different task. In the WWTP domain, CBR has been used for designing most suitable operations for a set of determined input contaminants.

In [4] was presented a CBR system that is part of a target architecture named DAI-DEPUR, which is able to support reasoning in a poorly understood and ill-structured domain and to learn from previously solved problems and to adapt the available experiential knowledge over the domain (dynamic learning environment). Additional works using CBR for environmental engineering problems appear in [5] and [6]. In the last one, a comparative study of the use of similarity measures was done.

In [7] the learning system that generates several expert system rules for treatment technologies from the treatability database was presented. It was used the ID3 machine learning algorithm to learn from a set of samples and produce humanly readably rules with relative sense.

Paper [8] talks about five different machine learning and statistical algorithms applied to the task of classifying descriptions of the daily operation of a Wastewater Treatment Plant, for instance CN2 and C4.5 give good accuracy.

3. ANN-CBR Hybrid model

In order to solve the problem a hybrid model was applied [9]. This model combines CBR and ANN along with fuzzy sets. This model takes into account the existence of values or neurons that exemplify the ANN construction. The use of fuzzy sets enables us to select neurons and enhance precision.

In this case a multi-layer net is used, having a layer (group) for each feature. In each group a node for each element of the dominance of the corresponding field is located. A non-directed arch exists between all the pairs of nodes except between the ones located in the same group.

For each arch in ANN a weight W_{ij} exists. Usually the learning procedure of the ANN is the one responsible for determining the weight associated with each arch. The case base described above is taken as the training set; each article of this case base will be an example.

The cases-based explanation consists of justifying the solution given to the problem by the ANN presenting the similar cases to the problem and their solution. The similarity can be calculated using the similarity function.

This comparison function has the interval [0, 1] as an image. It reaches value l when two sets of values for a predicting feature has the same force for the prediction of the value for the objective feature, and reaches value 0 when one of the set of values has a maximum predicting force and the other one, none. The comparison between the value in the problem P and case R is based on the relation of these values with the value inferred to the goal feature.

4. Hybrid model applied to wastewater treatment

In order to apply the selected model to wastewater treatment, a cases base should be defined. In problem solving, experts consider the characteristics of residual waters that are listed below:

- Biological oxygen demand (Low, Slightly Low, Medium, High)
- Chemical Oxygen Demand (Low, High)
- BOD/COD rate (Low, Medium, High)
- Water's acidity once it leaves the plant (Acid, Neutral, Basic)
- Nitrogen compounds (Low, High)
- Phosphorous compounds (Low, High)
- The residue after the evaporation (Low, High)
- Volatile Total Solids (Low, High)
- Filterable Solids (Low, High)
- Volatile Dissolved Solids (Low, High)
- Suspended Solids (Low, High)
- Settleable Solids (Low, High)

They are considered predicting features of the problem, having continuous values. To the effects of decision making, specified categories are considered after obtaining them via a discretization method [10], i. e., pH is taken as acid, neutral or basic as categories, not by its values.

The objective feature refers to the treatment to carry out, which is specified by means of a treatment outline. A treatment outline is a sequence of stages of the process. The possible values of stages are: mixed and neutralization, anaerobic lagoon, facultative lagoon, aerobic lagoon, fertile irrigation, coagulation and flocculation, sedimentation, oxidation/reduction, activated sludge and sifting.

Specifying a treatment outline is a task that requires expert knowledge. Human experts have to select the appropriate stages and the order in which they are going to be carried out. Therefore for each problem in this feature, the sequence of stages is specified to suit the outline of appropriate treatment for the case.

In an initial study of this problem, 29 representative cases of 3 types of Cuban industries were considered. The absence of information, which accounts for 40.22% of the data, is a characteristic to be considered when selecting the appropriated solution method.

In the topology of ANN, a group corresponds with a characteristic specified above. Each group has a node for each corresponding category. In the calculation of W_{ij} of the network's connections, the Pearson's correlation coefficient [9] was used.

The ANN's output would be a partial solution to the problem, because it will considerate the stages of the process to be included in the treatment outline, but not the execution order of these stages. In such a way, the ANN is complemented with the cases-based component which retrieves similar cases (past solutions) to the present problem. The user can estimate, in the context of the solutions given to these similar problems, the correct order of the treatment stages proposed by the ANN.

5. Conclusions

It is feasible to apply the proposed model to this problem because:

- It allows information absence. Information is taken from cases that contribute to train the ANN.
- The objective feature may take multiple values at once. Other neural network models, like MLP allow inferring a single value.
- The design of the similarity function doesn't require an additional effort, because it uses the weights of the ANN.
- Handling of continuous features is natural by means of categories. To each neuron in the ANN, we associate one of the possible categories, and it even allows their handling in a crisp and fuzzy way.

The results obtained with the ANN to propose the treatment to carry out were satisfactory (prediction accuracy of 90%). Besides according to the expert points of view, the application of this model is feasible. It is possible to obtain the appropriate treatment from the union of ANN's output and the most similar past cases.

Acknowledgments

The authors wish to thank VLIR (Vlaamse InterUniversitaire Raad, Flemish Interuniversity Council, Belgium) for supporting this work under the IUC Program VLIR-UCLV.

References

[1] J.L. Bueno, H. Sastre y A.G. Lavin, Contamination and Environmental Engineering. Volume III: Water Contamination. Edition FICYT (1997).

[2] Amhed Gamal The-DIN, Daniel W. Smith and Mohamed Gamal The-DIN, neural Application of artificial networks in wastewater treatment, Journal of Environmental Engineering and Science, supplement S1, pages S81-S95 (2004).

[3] KP Oliveira-Esquerre, M Mori, RE Bruns, Simulation of an industrialist to wastewater treatment plant using artificial neural networks and main components analysis, BRAZ J CHEM ENG 19(4): 365-370 (2002).

[4] M. Sánchez-Marré, U. Cortés, I.R.-Roda, M. Poch, and J. Lafuente. Learning and Adaptation in WWTP through Case-Based Reasoning. Special issue on Machine Learning, Microcomputers in Civil Engineering 12(4), 251-266 (1997).

[5] I.R. Rodriguez-Roda, M. Sánchez-Marré, J. Comas, J. Baeza, J. Colprim, J. Lafuente, U. Cortés and M. Poch, A hybrid supervisory system to support WWTP operation: implementation and validation, Water Science & Technology Vol 45 No 4-5 pp 289–297 (2002).

[6] Héctor Núñez, Miquel Sánchez-Marré, Ulises Cortés, Joaquim Comas, Montse Martínez, Ignasi Rodríguez-Roda and Manel Poch, A comparative study on the use of similarity measures in case-based reasoning to improve the classification of environmental system situations, Environmental Modelling & Software, 19(9), 809-819 (2004).

[7] Krovvidy, S (1998): Intelligent for Tools to wastewater treatment design. Civil Computer-Aided and Infrastructure Engineering, volume 13, pp. 219-226; ISSN: 1093-9687 (1998).

[8] J Commas, S Dzeroski, K Gibert, TO GO Stem, M Sánchez-Marré, Knowledge discovery by means of inductive methods in wastewater treatment plant dates, AI COMMUNICATIONS, 14(1): 45-62 (2001).

[9] María Matilde García Lorenzo, Yanet Rodríguez Sarabia, Rafael Esteban Bello Perez: Usando conjuntos borrosos para implementar un modelo para sistemas basados en casos interpretativos, In Proceedings of IBERAMIA-SBIA. Eds por M. C. Monard y J.S. Sichman, Sao Paulo, Brasil, November 19-22, 2000, pp. 1-11 (2000).

[10] H. Liu and R. Setiono, Chi2: Feature Selection and Discretization of Numeric Attributes, Proceedings of the IEEE 7th International Conference on Tools with Artificial Intelligence, Washington the USA, pp. 388-391 (1995).

[11] Núria Vidal, René Banares-Alcántara, Ignasi Rodríguez-Roda, and Manel Poch, Design of Wastewater Treatment Plants Using a Conceptual Design Methodology, Industrial & Engineering Chemistry Research, 41(20), 4993- 5005 (2002).

[12] M. Fiter, J. Colprim, Ms. Poch, Is. Rodríguez-Roda, "Enhancing biological nitrogen removal in a small WWTP by regulating the air supply" Water Sci Technol., 48(11-12), 445 - 452. ISSN: 0273-1223 (2003).

Brill Academic Publishers
P.O. Box 9000, 2300 PA Leiden.
The Netherlands

Lecture Series on Computer and Computational Sciences
Volume 2, 2005, pp. 136-140

Resource Packing to Fit Grid Infrastructure

Yu Xiong and Daizhong Su*

Advanced Design and Manufacturing Engineering Centre, SBE,
Nottingham Trent University, Burton Street, Nottingham, NG1 4BU, UK.
* Corresponding author, Professor of Design Engineering, E-mail: daizhong.su@ntu.ac.uk

Received 15 April, 2005; accepted in revised form 26 April, 2005

Abstract: Although Grid Computing has significant advantages, it still has not gained applications in practice as widely as expected, due to the problem of utilizing existing resources. To overcome the problem, the latest Grid architecture, *Web Service Resource Framework*, has been applied and a method has been developed for packing existing resources, including different types of software, hard disk spaces and other devices, to fit the Grid architecture.

Keywords: Grid, WSRF, Distributed Computing.

1. Introduction.

Grid computing is an emerging technology, which enables the utilization of distributed computing and resources to create a single system image, granting users seamless access to vast IT capabilities. Just like an Internet user views a unified instance of content via the Web, a Grid user essentially sees a single, large virtual computer. The concept of virtual organization is strengthened in Grid, which has the potential to change dramatically the way in which we use computers to resolve problems, similar to that Web has changed the way we exchange information [1,2].

Although Grid has significant advantages, it still has not gained applications in practice as widely as expected, due to the problem of utilizing existing resources. Many resources, including both standalone and web-enabled applications, currently in use are not initially designed for Grid. Users or designers of these applications/resources are reluctantly to rebuild them to adopt the new structure, because the rebuilding could be expensive in time and cost..To overcome the problem, in this research, the latest Grid architecture, *Web Service Resource Framework* (WSRF) [4], has been applied and a method has been developed for packing existing resources to fit the Grid architecture.

2. Converting Resources into Grid Supported Services within WSRF

WSRF is a set of Web service specifications and conventions designed to standardize their representation. In this research, it is utilized to coordinate Web service and Grid. Different from Grid, Web service is a mature and practically-proven technology. With Web service, the request and the response are encoded in XML. An XML formatted remote procedure call is sent and received by the *XML-Remote Process Call* function. The *Simple Object Access Protocol*, which is also an XML technology, is utilised to remotely manipulate an object. The existences are published through a number of different Web service protocols, including *Universal Description*, *Discovery and Integration* and *Web Services Description Language*, which enables the Grid computing to be easy to deploy [3].

Resources play an important role in distributed systems. But majority of existing resources are not ready for Grid, an emerging new technology. While most of programs are following the Object-Oriented structure that is still very popular, the Service-Oriented ones would have not got a sufficient market for lacking of compatible resources.

In this research, a method has been developed to pack existing applications with three types of services including: (1) *Application proxy service (APS)* which provides a surrogate to control the access to a shared application. It models functions presented in a shared application (SA) and exposes the functions as services. Actually, an APS is similar to a running SA process and can be accessed remotely. The APS also manages the resource used by the shared application, monitors the situation of the shared application, and handles security and other issues. A client application will interact with an APS to share the application. (2) *Application proxy factory service (APFS).* Some of Applications may not be used in a multi-user environment (a service-oriented software should be a multi-user environment). A single APS cannot serve multiple users, so there will be multiple services to serve different clients. These APSs should be managed, and the resources used by these services should be managed too. The APFS, which is similar to the process management in local operation system, is designed to meet the needs of managing the creation and lifecycle of an APS. The resources used by these services are also managed by the APFS. Each shared application has an APFS to create multiple APS instances. The APS and APFS patterns not only make it possible to pack an application with a web service, but also enable a single user application to support multiple users. (3) <u>Application manager service (AMS).</u> In an Internet environment, everyone can access the same web service from any client end at the same time. Although APS and APFS are used to make a shared application pretend to be a web service for multi-users, it is in fact a single user one. The number of clients that these applications can serve at the same time are not as many as that of a pure Web service. So AMS is built to confront this performance issue. Another reason of the building application manager is to support the access control mechanism.

Figure 1 shows the architecture of the approach, there may be multiple shared applications installed on multiple computers. Each of them is installed with the APFS. The running process of the shared application is represented by the APS created and managed by the APFS. All the APFSs of a shared application should be registered to the AMS and assigned an application identifier. The AMS has an *Access Control List* to control the security.

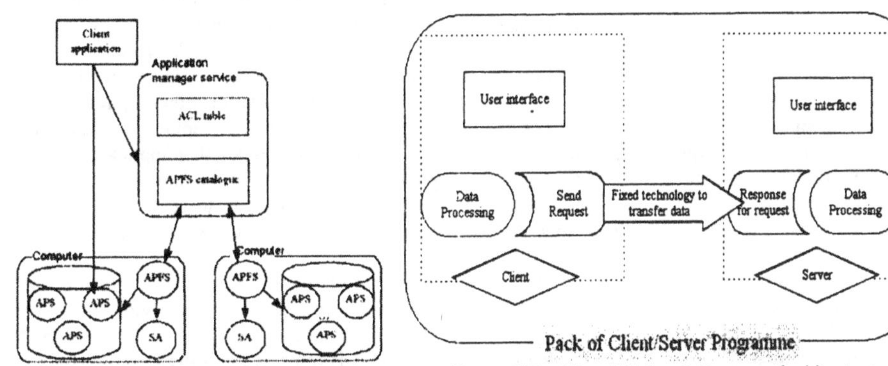

Figure 1: Architecture of the approach

Figure 2 Traditional Network Program Architecture

From the client's view, when the user launches a client application, the client application will issue a request to get an APS from the AMS. When the request arrives, the AMS will look up the APFS registered in the catalogue using the application identifier specified by the client application. If the APFS exists in the catalogue, the AMS will check whether the client application has the right to use it. If the access right checking were successful, the AMS would interact with the APFS to create an APS.

There may be multiple APFS associated with a shared application. These services are installed in different host for performance reasons. When an AMS sends a request to an APFS to create an APS, the APFS should determine whether a new APS could be created or not. If this APFS can't create a new APS, the AMS will be informed, then the AMS will choose another APFS to try.

After an APS has been created by an APFS, the handle of the newly created APS will be returned to the client application by the AMS. The AMS authorizes the client application with a license to access the APS. Then the client application will use the handle to access the shared application. After the client application completes its task, the APS used by the client application will be destroyed, and the resources allocated to the APS will be released.

During the process of interaction described above, the client application cannot access the shared application without authorization. The performance issue can be resolved by AMS and multiple APFSs. The service-oriented software sharing can thus be achieved.

3. Resource Packing.

The method mentioned in the above section is applied to pack different types of resources as described bellow.

<u>A. Standalone software packing.</u>

Three kinds of standalone applications are considered:

(1). Applications built with component technology

If all the business logics of an application reside in a single component, the solution is to simply pack the component object and expose the web service interface. If the business logic resides in multi-components, a new business object that interacts with the multi-components and exposes all of business logic interfaces should be built. Then the business object can be packed with the APS.

(2). Applications using shared libraries

In this case, the adapter is built first to call functions within the shared libraries, a high-level business object is then built to interact with low-level adapters, and, finally, the business interface is exposed with the APS. In some situations, shared libraries can be converted into components, and then this method can be used to describe it.

(3). Standalone executable files

This kind of applications is the most difficult for packing. Although it is possible to convert these kinds of applications into a web service, it is not worth doing so in most situations because of the difficulties. Ideally, rewriting the application could be a better choice. However some applications may be important and cost a high price to rewrite, therefore, the conversion has to be considered with different cases described below:

Some of these applications can be converted into a shared library by tools, for example, Microsoft Visual Studio .Net provides some tools to convert standalone executable files into *Dynamic Link Library*. In this case, the previously packed shared libraries with APS policy can be used.

Some applications use input stream to read data and output stream to write results, such as applications using standard I/O stream. In many platforms and programming environments, there are functions which can replace or redirect the input stream or output stream with another input stream or output stream. For example, Java language can use System.setIn() method to replace standard input stream with any other types of input streams, even network input stream. So another application can be written to replace or redirect the I/O stream of the application, and to expose the interface using Web service. Other cases are the applications that use command line arguments to read data. In many programming languages, there are functions which can execute a standalone executable file and specify the command line arguments.

The remaining applications (most of them are *Graphical User Interface* applications) can neither use standard I/O functions nor be converted into shared libraries. To reuse such applications instead of rewriting them, it often results in a high cost. In order to reuse these applications, a Bypass library should be written. Then the library needed by the applications is replaced with the Bypass one. The Bypass library will interact with the APS. As can be seen, this method will incur high cost and not be efficient, therefore, it is not recommended.

B. Network enabled software packing

There are applications which are already network-supported, however, they are developed with different technologies which are not service-oriented. Such applications cannot be viewed as a single programme as the standalone one does.

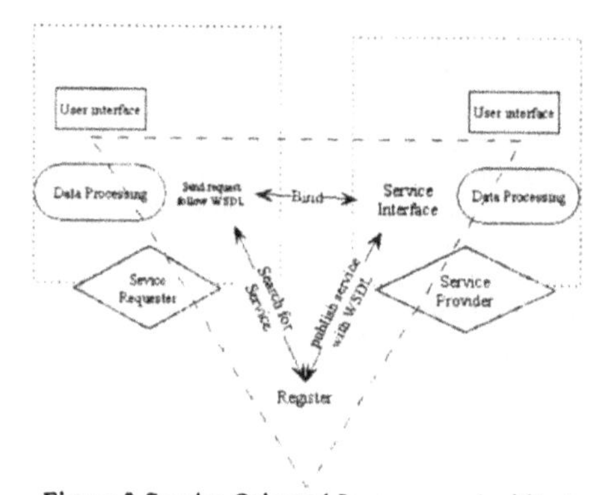

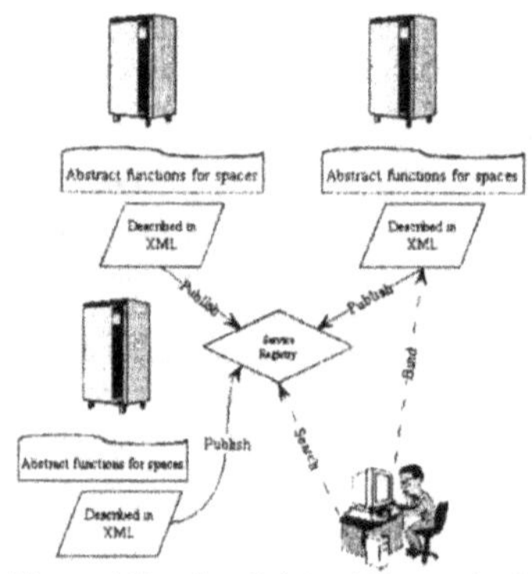

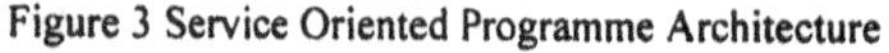
Figure 3 Service Oriented Programme Architecture

Figure 4 Service-Oriented Space sharing

Most of the applications have two modules in a traditional view: client and server, as shown in Figure 2, and both of the two sides have to programme with compatible technology. The client and server have to be born in the same time, people cannot code the client program if they do not know the server code well. Usually, such program is coded by the same group of people for both of the two sides. However, to change it into service oriented, part of the code does need to be re-coded.

As shown in Figure 3, the structure of the applications includes the service requester module, the service provider module and the middleware module (sending request, service interface and register within the triangle area). This helps to change it into a Service-oriented application. No matter what the program language or technology has been used. Only the request and response functions are to be replaced. They can be rewritten following the service way, but the core code of data processing will still be reserved to work in the new model. Following this structure, multi-user access can be achieved.

C. Hard disk space and other resources packing

Hard disk space can be shared in network as a service, and its disk functions can be operated as well, such as *copy, paste, view property, sharing limitations* and even *delete* according to the authorization level. Participant, who wants to share his hard disk space to others via the service, needs to write a program in the format suitable for the service including the disk operation functions mentioned above. These functions are described in XML and published on the service registry as mentioned in Section 2. At the beginning, the user may do not know the location of the space provider; however, after having found the XML service description file in service registry, he is redirected to and bound with the provider, and is able to operate the shared spaces with the functions provided. The GridFTP is utilized in this research, which provided a high-performance, secure, reliable data transfer protocol optimized for high-bandwidth wide-area networks [5].

As shown in Figure 4, the requester, i.e., the user, may not be aware of which provider he is bound to. According to his request, he could be bound with two or even more providers, but what appears to him, he is just using his own computer and find a space which seems located in his own hard disk.

Just the same as the utilisation of the hard disk space, any other resources can be represented as applications and converted to be services in the Grid environment. Their functions are abstracted out first, packed as described in Section 2, and then published in service registry so that can be found by users.

All the situations described above can be packed using the application manager service, application proxy factory service and application proxy service models. It requires a stateful service, which is presented by WSRF [4]. Take the application proxy service for example, each application proxy service should be associated with one active process of shared application, and also invoked by only one client application. If the application proxy service is stateless, the aim can't be achieved. The same situation happens in the implementation of the application proxy factory service.

4. Conclusion

To utilize Grid computing in a more practical way and to enrich the Grid resources, a method of converting traditional resources to services has been successfully developed. It can be used to pack difference resources to fit the Grid architecture, including the packing for standalone software, network enabled software, hard disk space and other devices.

The outcome of this research has greatly enhanced the usability of Grid, and proved that existing traditional resources can be re-used for Grid with the method developed and compatible with most other advanced techniques.

Further research to be conducted will include a universal Grid portal that can monitor and invoke the service resources.

Acknowledgement

The authors are grateful for the support received from the EU Asia-Link programme (grant No.ASI/B7-301/98/679-023) and Asia IT&C programme (Grant No. ASI/B7-301/3152-099/71553).

References

[1] I. Foster, C. Kesselman and S. Tuecke, '*The Anatomy of the Grid: Enabling Scalable Virtual Organizations*, http://www.globus.org/research/papers/anatomy.pdf, accessed 20th Dec, 2004

[2] Karl Czajkowski, Don Ferguson, Ian Foster, Jeff Frey, Steve Graham, Tom Maguire, David Snelling, and Steve Tuecke , *Infrastructure to WS-Resource Framework: Refactoring and Extension*, globus project whitepaper, http://www.globus.org/wsrf/specs/ogsi_to_wsrf_1.0.pdf, Accessed 10th Dec, 2004

[3] M. C. Brown, "Merging Grid and Web services", IBM DeveloperWorks, http://www-128.ibm.com/developerworks/grid/library/gr-web/, accessed 12th ,Dec. 2004
[4] K. Czajkowski, D. Ferguson and I. Foster, *'The WS-Resource Framework*, globus project whitepaper, http://www.globus.org/wsrf/specs/ws-wsrf.pdf, Accessed on 10th Dec, 2004
[5] V. Silva, *Transferring files with GridFTP*, IBM DeveloperWorks, http://www-106.ibm.com/developerworks/grid/library/gr-ftp/, accessed 5th March 2005.

Brill Academic Publishers
P.O. Box 9000, 2300 PA Leiden,
The Netherlands

Lecture Series on Computer and Computational Sciences
Volume 2, 2005, pp. 141-145

A Procedure Based on Synthetic Projection Model for Multiple Attribute Decision Making in Uncertain Setting

Z. S. Xu[1]

College of Economics and Management, Southeast University,
Nanjing, Jiangsu 210096, China

Received 5 March, 2005; accepted in revised form 16 March, 2005

Abstract: The aim of this paper is to investigate the multiple attribute decision making problem with preference information on alternatives in uncertain setting, in which the information about attribute weights is partly known, and the attribute values and the decision maker's preference values are expressed in the form of interval numbers. We establish the lower bound projection model, the upper bound projection model and their synthetic projection model respectively. By solving the synthetic projection model, we can determine the vector of attribute weights. Finally, a procedure is suggested for multiple attribute decision making under uncertainty, and then, the developed procedure is applied to determine the most desirable investment alternative of an investment company.

Keywords: Multiple attribute decision making; projection model; preference; interval number.
Mathematics Subject Classification: 90D35, 65F15.

1. Introduction

Multiple attribute decision making involves finding the most desirable alternative(s) from a given alternative set. In general, the decision information can only be obtained incompletely because of time pressure, lack of knowledge or data, and the decision maker's limited expertise related with problem domain [1]. Many decision making processes, in the real life, take place in situations in which the infotmation on attribute weights are not precisely known, but value ranges can be obtained. Up to now, some methods [2-7] have been proposed to deal with these problems. However, all these approaches will fail in dealing with the situations in which not only the attribute weights and attribute values take the forms of interval numbers, but also the decision maker has preferences for alternatives (we call these problems UMADM-PIA, for short). The aim of this paper is to develop a procedure for UMADM-PIA. In order to do so, we first establish three projection models called the lower bound projection model, the upper bound projection model and their synthetic projection model respectively. Then we determine the attribute weights by solving the synthetic projection model, and present an algorithm for ranking all the alternatives and selecting the most desirable one(s).

2. Main Results

For convenience, we let $M = \{1,2,\dots,m\}$ and $N = \{1,2,\dots,n\}$. For a multiple

[1] Corresponding author. E-mail: xu_zeshui@263.net

attribute decision making problem, let $X=\{x_1,x_2,...,x_n\}$ be the finite set of alternatives, $U=\{u_1,u_2,...,u_m\}$ be the finite set of attributes, and $w=(w_1,w_2,...,w_m)^T$ be the vector of attribute weights, $w \in H$, where H is the set of the known weight information. The decision maker also has a preference for alternative $x_i \in X$, and let the preference value be $\tilde{v}_i$, where $\tilde{v}_i=[v_i^L,v_i^U]$, $0 \le v_i^L \le v_i^U \le 1$. Let $\tilde{A}=(\tilde{a}_{ij})_{n\times m}$ be the decision matrix, where $\tilde{a}_{ij}=[a_{ij}^L,a_{ij}^U]$ is an attribute value, which takes the form of interval number, for alternative x_i with respect to attribute u_j. In general, there are benefit attributes and cost attributes in multiple attribute problems. In order to measure all attributes in dimensionless units, we can normalize each attribute value $\tilde{a}_{ij}$ in matrix $\tilde{A}=(\tilde{a}_{ij})_{n\times m}$ into a corresponding element in matrix $\tilde{R}=(\tilde{r}_{ij})_{n\times m}$ using the following formulas:

$$\tilde{r}_{ij}=\tilde{a}_{ij}\Big/\sum_{i=1}^{n}\tilde{a}_{ij}, \quad \textit{for benefit attribute } u_j, \ i\in N, j\in M \tag{1}$$

$$\tilde{r}_{ij}=\left(1/\tilde{a}_{ij}\right)\Big/\sum_{i=1}^{n}\left(1/\tilde{a}_{ij}\right), \quad \textit{for benefit attribute } u_j, \ i\in N, j\in M \tag{2}$$

By the operations of interval numbers, we rewrite (1) and (2) as (3) and (4) respectively:

$$\begin{cases} r_{ij}^L = a_{ij}^L\Big/\sum_{i=1}^{n} a_{ij}^U \\ r_{ij}^U = a_{ij}^U\Big/\sum_{i=1}^{n} a_{ij}^L \end{cases}, \quad \textit{for benefit attribute } u_j, \ i\in N, j\in M \tag{3}$$

$$\begin{cases} r_{ij}^L = \left(1/a_{ij}^U\right)\Big/\sum_{i=1}^{n}\left(1/a_{ij}^L\right) \\ r_{ij}^U = \left(1/a_{ij}^L\right)\Big/\sum_{i=1}^{n}\left(1/a_{ij}^U\right) \end{cases}, \quad \textit{for benefit attribute } u_j, \ i\in N, j\in M \tag{4}$$

By the normalized decision matrix $\tilde{R}=(\tilde{r}_{ij})_{n\times m}$ and the vector $w=(w_1,w_2,...,w_m)^T$, we can get the aggregated value of alternative x_i as follows:

$$\tilde{z}_i(w)=\sum_{j=1}^{m}\tilde{r}_{ij}w_j, \quad i\in N \tag{5}$$

Definition 1[8]. Let $\tilde{a}=[a^L,a^U]$ and $\tilde{b}=[b^L,b^U]$, and let $l_{\tilde{a}}=a^U-a^L$ and $l_{\tilde{b}}=b^U-b^L$, then the degree of possibility of $\tilde{a}\ge\tilde{b}$ is defined as:

$$p(\tilde{a}\ge\tilde{b})=\max\left\{1-\max\left((b^U-a^L)/(l_{\tilde{a}}+l_{\tilde{b}}),0\right),0\right\} \tag{6}$$

Obviously, from Definition 1, we know that $p(\tilde{a}\ge\tilde{b})$ satisfies the complementary property, i.e.,

$$0\le p(\tilde{a}\ge\tilde{b})\le 1, \quad p(\tilde{a}\ge\tilde{b})+p(\tilde{b}\ge\tilde{a})=1, \quad p(\tilde{a}\ge\tilde{a})=0.5 \tag{7}$$

In the following, we first give some notations:

Let $\tilde{z}(w)=(\tilde{z}_1(w),\tilde{z}_2(w),...,\tilde{z}_n(w))^T$, where $\tilde{z}_i(w)=[z_i^L(w),z_i^U(w)]=\sum_{j=1}^{m}\tilde{r}_{ij}w_j=$ $[\sum_{j=1}^{m}r_{ij}^L w_j,\sum_{j=1}^{m}r_{ij}^U w_j]$, and let $z^L(w)=(z_1^L(w),z_2^L(w),...,z_n^L(w))^T$, $z^U(w)=(z_1^U(w),z_2^U(w),...,z_n^U(w))^T$, $\tilde{z}(w)=[z^L(w),z^U(w)]$. Let $\tilde{v}=(\tilde{v}_1,\tilde{v}_2,...,\tilde{v}_n)^T$, where $\tilde{v}=[v^L,v^U]$, $v^L=(v_1^L,v_2^L,...,v_n^L)^T$, $v^U=(v_1^U,v_2^U,...,v_n^U)^T$. Usually, if the vector of aggregated values, $\tilde{z}(w)$, is the same as the vector of preference values, $\tilde{v}$, then we can utilize the vector of preference values, $\tilde{v}$, to rank the alternatives and select the most desirable one(s). However, in the real life, there always exists a difference between the vector of aggregated values, $\tilde{z}(w)$, and the vector of preference values, $\tilde{v}$. To determine the vector of weight attributes, we can minimize the difference between the vector of the aggregated values, $\tilde{z}(w)$, and the vector of preference values, $\tilde{v}$. Therefore, we set

$$\cos\theta_1=\cos<z^L(w),v^L>=\frac{\sum_{i=1}^{n}z_i^L(w)v_i^L}{\sqrt{\sum_{i=1}^{n}\left(z_i^L(w)\right)^2}\sqrt{\sum_{i=1}^{n}\left(v_i^L\right)^2}} \tag{8}$$

$$\cos\theta_2=\cos<z^U(w),v^U>=\frac{\sum_{i=1}^{n}z_i^U(w)v_i^U}{\sqrt{\sum_{i=1}^{n}\left(z_i^U(w)\right)^2}\sqrt{\sum_{i=1}^{n}\left(v_i^U\right)^2}} \tag{9}$$

Obviously, the greater the values of $\cos\theta_1$ and $\cos\theta_2$, the closer the direction of $z^L(w)$ to that of v^L, and the closer the direction of $z^U(w)$ to that of v^U. As it is well known that a vector is composed of direction and modular size, $\cos\theta_1$ and $\cos\theta_2$, however, only reflect the similarity measures of the directions of the vectors $z^L(w)$ and v^L, $z^U(w)$ and v^U respectively. In order to measure the similarity degree between vectors $z^L(w)$ and v^L, and the similarity degree between vectors $z^U(w)$ and v^U respectively. In the following we introduce the formulas of projection of the vector $z^L(w)$ on the vector v^L, and of projection of the vector $z^U(w)$ on the vector v^U as follows respectively:

$$\Pr j_{v^L}\left(z^L(w)\right)=\left|z^L(w)\right|\cos\theta_1=\sqrt{\sum_{i=1}^{n}\left(z_i^L(w)\right)^2}\frac{\sum_{i=1}^{n}z_i^L(w)v_i^L}{\sqrt{\sum_{i=1}^{n}\left(z_i^L(w)\right)^2}\sqrt{\sum_{i=1}^{n}\left(v_i^L\right)^2}}=\frac{\sum_{i=1}^{n}z_i^L(w)v_i^L}{\sqrt{\sum_{i=1}^{n}\left(v_i^L\right)^2}}$$

$$=\sum_{i=1}^{n}z_i^L(w)\bar{v}_i^L \tag{10}$$

Similarly,

$$\Pr j_{v^U}\left(z^U(w)\right)=\left|z^U(w)\right|\cos\theta_2=\sum_{i=1}^{n} z_i^U(w)\bar{v}_i^U \tag{11}$$

where

$$\left|z^L(w)\right|=\sqrt{\sum_{i=1}^{n}\left(z_i^L(w)\right)^2},\quad \left|z^U(w)\right|=\sqrt{\sum_{i=1}^{n}\left(z_i^U(w)\right)^2}$$

are the modules of $z^L(w)$ and $z^U(w)$ respectively, and

$$\bar{v}_i^L=\frac{v_i^L}{\sqrt{\sum_{i=1}^{n}\left(v_i^L\right)^2}},\quad \bar{v}_i^U=\frac{v_i^U}{\sqrt{\sum_{i=1}^{n}\left(v_i^U\right)^2}}.$$

Obviously, the greater the values $\Pr j_{v^L}\left(z^L(w)\right)$ and $\Pr j_{v^U}\left(z^U(w)\right)$, the closer $z^L(w)$ to v^L, and the closer $z^U(w)$ to v^U, i.e., the closer $\tilde{z}(w)$ to $\tilde{v}$. Therefore, we establish the lower bound projection model (LPM) and the upper bound projection model (UPM) as follows respectively:

(LPM)

$$\max \Pr j_{v^L}\left(z^L(w)\right)=\sum_{i=1}^{n} z_i^L(w)\bar{v}_i^L$$

$$s.t.\quad w\in H$$

(UPM)

$$\max \Pr j_{v^U}\left(z^U(w)\right)=\sum_{i=1}^{n} z_i^U(w)\bar{v}_i^U$$

$$s.t.\quad w\in H$$

To obtain a unique vector of attribute weights. $w=(w_1,w_2,\ldots,w_m)^T$, we combine the models (LPM) and UPM into the following synthetic projection model (SPM):

(SPM)

$$\max \Pr j_{v}\left(z(w)\right)=\sum_{i=1}^{n}\left(z_i^L(w)\bar{v}_i^L+z_i^U(w)\bar{v}_i^U\right)$$

$$s.t.\quad w\in H$$

By solving this model, we obtain the optimal solution $w^*=(w_1^*,w_2^*,\ldots,w_m^*)^T$, and then by (5), we get the aggregated value $\tilde{z}_i(w^*)(i\in N)$. In order to rank all alternatives, we utilize (6) to compare each two interval numbers $\tilde{z}_i(w^*)$ and $\tilde{z}_j(w^*)$ $(i,j\in N)$, and establish the possibility degree matrix $P=(p_{ij})_{n\times n}$, where

$$p_{ij}=P(\tilde{z}_i(w^*)\geq\tilde{z}_j(w^*)),\quad 0\leq p_{ij}\leq 1,\quad p_{ij}+p_{ji}=1,\quad p_{ii}=0.5 \tag{12}$$

By Definition 1, we know that P is a fuzzy complementary matrix [8-14]. Summing all elements in each line of matrix P, we have

$$p_i=\sum_{j=1}^{n} p_{ij},\quad i\in N \tag{13}$$

Then we can rank all alternatives using the value of p_i, $i\in N$, and get the most desirable one(s).

Acknowledgements

The author wishes to thank the anonymous referees for their careful reading of the manuscript and their fruitful comments and suggestions. This work was supported by China Postdoctoral Science Foundation (2003034366).

References

[0] S.H. Kim and B.S. Ahn, Interactive group decision making procedure under incomplete information, *European Journal of Operational Research* **116** 498-506(1999).

[0] K.S. Park, S.H. Kim and Y.C. Yoon, Establishing strict dominance between alternatives with special type of incomplete information, *European Journal of Operational Research* **96** 398-406(1997).

[0] N. Bryson and A. Mobolurin, An action learning evaluation procedure for multiple criteria decision making problems, *European Journal of Operational Research* **96** 379-386(1997).

[0] K.S. Park and S.H. Kim, Tools for interactive multi-attribute decision making with incompletely identified information, *European Journal of Operational Research* **98** 111-123(1997).

[0] S.H. Kim, S.H. Choi and J.K Kim, An interactive procedure for multiple attribute group decision making with incomplete information: Range-based approach, *European Journal of Operational Research* **118** 139-152(1999).

[0] Q.L. Da and Z.S. Xu, Single objective optimization model in uncertain multi-attribute decision-making, *Journal of Systems Engineering* **17** 50-55(2002).

[0] Z.S. Xu: *Study on Methods for Multiple Attribute Decision Making under Some Situations*. Ph.D thesis, Southeast University, Nanjing, China, 2002.

[0] Z.S. Xu and Q.L. Da, The uncertain OWA operator, *International Journal of Intelligent Systems* **17** 569-575(2002).

[0] H. Nurmi, Approaches to collective decision making with fuzzy preference relations, *Fuzzy Sets and Systems* **6** 249-259(1981).

[0] F. Chiclana, F. Herrera and E. Herrera-Viedma, Integrating multiplicative preference relations in a multipurpose decision-making based on fuzzy preference relations, *Fuzzy Sets and Systems* **122** 277-291 (2001).

[0] Z.S. Xu and Q.L. Da, An approach to improving consistency of fuzzy preference matrix, *Fuzzy Optimization and Decision Making* **2** 3-12(2003).

[0] Z.S. Xu, Goal programming models for obtaining the priority vector of incomplete fuzzy preference relation, *International Journal of Approximate Reasoning* **36** 261-270(2004).

[0] Z.S. Xu: *Uncertain Multiple Attribute Decision Making: Methods and Applications*. Tsinghua University Press, Beijing , 2004.

[1] Z.S. Xu and Q.L. Da, A least deviation method to obtain a priority vector of a fuzzy preference relation, *European Journal of Operational Research* **164** 206-216 (2005).

Brill Academic Publishers
P.O. Box 9000. 2300 PA Leiden,
The Netherlands

Lecture Series on Computer and Computational Sciences
Volume 2, 2005, pp. 146-149

A Method for Fuzzy Decision Making Problem with Triangular Fuzzy Complementary Judgment Matrix

Z. S. Xu[1]

College of Economics and Management, Southeast University,
Nanjing, Jiangsu 210096, China

Received 5 March, 2005; accepted in revised form 16 March, 2005

Abstract: The aim of this paper is to investigate the fuzzy decision making problem, in which the decision information takes the form of triangular fuzzy complementary judgment matrix. Some new concepts such as triangular fuzzy complementary judgment matrix, modal value complementary judgment matrix, left deviation matrix and right deviation matrix, etc., are defined. Based on error analysis, a priority method for triangular fuzzy complementary judgment matrix is proposed. The decision alternatives are ranked by the priority formula of triangular fuzzy numbers.

Keywords: Triangular fuzzy complementary judgment matrix: error analysis; priority.
Mathematics Subject Classification: 90D35, 65F15.

1. Introduction

Multiple attribute decision making has received a great deal of attention from researchers in many disciplines [1-6]. In the process of decision making, a decision maker usually needs to compare a set of n alternatives with respect to a single criterion, and constructs a complementary judgment matrix [7-18] $A=(a_{ij})_{n\times n}$, whose element a_{ij} estimates the dominance of alternative i over j, and satisfies $a_{ij} \geq 0, a_{ij}+a_{ji}=1, a_{ii}=1/2$. However, many decision making processes, in the real world, take place in an environment in which the decision maker may have vague knowledge about the preference degree of one alternative over another, and can't estimate his/her preference with exact numerical value, but with interval number [19] or triangular fuzzy number. Therefore, it is necessary to pay attention to this issue. The aim of this paper is to investigate the fuzzy decision making problem with triangular fuzzy complementary judgment matrix. We define some concepts such as triangular fuzzy complementary judgment matrix, modal value complementary judgment matrix, left deviation matrix and right deviation matrix, etc. Then we propose a priority method based on error analysis for triangular fuzzy complementary judgment matrix. Furthermore, we develop an approach to ranking decision alternatives by the priority formula of triangular fuzzy numbers.

2. Method

For convenience, we let $\Omega=\{1,2,\ldots,n\}$.

Definition 1[7]. Let $A=(a_{ij})_{n\times n}$ be a judgment matrix, where $a_{ij} \geq 0$, $i,j \in \Omega$. If $a_{ij}+a_{ji}=c(c>0)$, $i,j\in\Omega$, then A is a complementary judgment matrix (or fuzzy

[1] Corresponding author. E-mail: xu_zeshui@263.net

preference relation).

Definition 2. Let $\widetilde{A}=(\widetilde{a}_{ij})_{n\times n}$ be a judgment matrix, where $\widetilde{a}_{ij}=[a_{ij}^{L},a_{ij}^{U}]$, $a_{ij}^{U}\geq a_{ij}^{L}\geq 0$, $i,j\in\Omega$. If $a_{ij}^{L}+a_{ji}^{U}=a_{ij}^{U}+a_{ji}^{L}=c\,(c>0)$, $i,j\in\Omega$, then we call $\widetilde{A}$ an interval complementary judgment matrix.

Definition 3. Let $\hat{A}=(\hat{a}_{ij})_{n\times n}$ be a judgment matrix, where $\hat{a}_{ij}=[a_{ij}^{L},a_{ij}^{M},a_{ij}^{U}]$, $a_{ij}^{U}\geq a_{ij}^{M}\geq a_{ij}^{L}\geq 0$, $i,j\in\Omega$. If

$$a_{ij}^{L}+a_{ji}^{U}=a_{ij}^{M}+a_{ji}^{M}=a_{ij}^{U}+a_{ji}^{L}=c,\ i,j\in\Omega$$

then we call $\hat{A}$ a triangular fuzzy complementary judgment matrix.

Definition 4. Let $\hat{A}=(\hat{a}_{ij})_{n\times n}$ be a triangular fuzzy complementary judgment matrix, then we call $A^{(M)}=(a_{ij}^{M})_{n\times n}$ the modal value complementary judgment matrix of $\hat{A}$.

Definition 5. Let $\hat{A}=(\hat{a}_{ij})_{n\times n}$ be a triangular fuzzy complementary judgment matrix, then we call $\delta^{(l)}=(\delta_{ij}^{l})_{n\times n}$ the left deviation matrix of $\hat{A}$, where $\delta_{ij}^{l}=a_{ij}^{M}-a_{ij}^{L}$, $i,j\in\Omega$.

Definition 6. Let $\hat{A}=(\hat{a}_{ij})_{n\times n}$ be a triangular fuzzy complementary judgment matrix, then we call $\delta^{(r)}=(\delta_{ij}^{r})_{n\times n}$ the right deviation matrix of $\hat{A}$, where $\delta_{ij}^{r}=a_{ij}^{U}-a_{ij}^{M}$, $i,j\in\Omega$.

Suppose that the decision maker utilize 0-1 scale [7-13] to construct a complementary judgment matrix $A=(a_{ij})_{n\times n}$, and let $w=(w_1,w_2,\ldots,w_n)^T$ be the priority vector of A, where $\sum_{i=1}^{n}w_i=1$, $w_i\geq 0$, $i\in\Omega$. In [11], Xu suggests a simple formula for obtaining the priority vector of a complementary judgment matrix as follows.

$$w_i=\frac{1}{n(n-1)}\left(\sum_{j=1}^{n}a_{ij}+\frac{n}{2}-1\right),\quad i\in\Omega \tag{1}$$

In the following we shall introduce the propagation of errors of a random function [20]:

Let $z=f(y_1,y_2,\ldots,y_n)$ be a random function, and the error of random variable $y_i\,(i\in\Omega)$ of the function z be $\sigma_{y_i}^{2}\,(i\in\Omega)$, then the propagation of errors of the function z is given by

$$\sigma_z^2=\sum_{i=1}^{n}\left(\frac{\partial f}{\partial y_i}\right)^2\sigma_{y_i}^2+2\sum_{1\leq i\leq j\leq n}\left(\frac{\partial f}{\partial y_i}\frac{\partial f}{\partial y_j}\rho_{ij}\sigma_{y_i}\sigma_{y_j}\right) \tag{2}$$

where ρ_{ij} is coefficient. If $\rho_{ij}=0$, then (2) can be simplified as [21]:

$$\sigma_z^2=\sum_{i=1}^{n}\left(\frac{\partial f}{\partial y_i}\right)^2\sigma_{y_i}^2 \tag{3}$$

In practical application, for the convenience of evaluation, (3) is usually replaced with the following formula:

$$(\Delta z)^2 = \sum_{i=1}^{n} \left(\frac{\partial f}{\partial y_i} \right)^2 (\Delta y_i)^2 \tag{4}$$

Based on the error analysis, in the following we shall propose a method for the fuzzy decision making problem with triangular fuzzy complementary judgment matrix.

Step 1. For a fuzzy decision making problem, let $X=\{x_1,x_2,...,x_n\}$ be a discrete set of alternatives. The decision maker compares each two alternatives with respect to a single criterion by using 0-1 scale, and constructs a triangular fuzzy complementary judgment matrix $\hat{A}=(\hat{a}_{ij})_{n\times n}$, where

$$\hat{a}_{ij}=[a_{ij}^L,a_{ij}^M,a_{ij}^U],\ a_{ij}^U \geq a_{ij}^M \geq a_{ij}^L \geq 0,\ a_{ij}^L+a_{ji}^U=a_{ij}^M+a_{ji}^M=a_{ij}^U+a_{ji}^L=1,\ i,j\in\Omega$$

Step 2. Calculate the left deviation matrix $\delta^{(l)}=(\delta_{ij}^l)_{n\times n}$ and the right deviation matrix $\delta^{(r)}=(\delta_{ij}^r)_{n\times n}$, where $\delta_{ij}^l=a_{ij}^M-a_{ij}^L$, $\delta_{ij}^r=a_{ij}^U-a_{ij}^M$, $i,j\in\Omega$.

Step 3. By (1), we calculate the priority vector $\overline{w}=(\overline{w}_1,\overline{w}_2,...,\overline{w}_n)^T$ of the modal value complementary judgment matrix $A^{(M)}=(a_{ij}^M)_{n\times n}$.

Step 4. By (1) and (2), we calculate the left maximal error vector $\Delta w^{(l)}=(\Delta w_1^{(l)}, \Delta w_2^{(l)},...,\Delta w_n^{(l)})^T$ and the right maximal error vector $\Delta w^{(r)}=(\Delta w_1^{(r)},\Delta w_2^{(r)},...,\Delta w_n^{(r)})^T$ corresponding to $\overline{w}$, where

$$\Delta w_i^{(l)}=\frac{1}{n}\sqrt{\sum_{j=1}^{n}(\delta_{ij}^l)^2},\quad \Delta w_i^{(r)}=\frac{1}{n}\sqrt{\sum_{j=1}^{n}(\delta_{ij}^r)^2},\quad i\in\Omega \tag{5}$$

and thus, we get the priority vector $\hat{w}=(\hat{w}_1,\hat{w}_2,...,\hat{w}_n)^T$ of $\hat{A}$, where

$$\hat{w}_i=[w_i^L,w_i^M,w_i^U]=[\overline{w}_i-\Delta w_i^{(l)},\overline{w}_i,\overline{w}_i+\Delta w_i^{(r)}],\quad i\in\Omega \tag{6}$$

Step 5. To rank all the alternatives, we calculate the expected value of the triangular fuzzy number $\hat{w}_i$ ($i\in\Omega$) by using the following formula [22]:

$$\hat{w}_i^{(\alpha)}=\frac{1}{2}[(1-\alpha)(\overline{w}_i-\Delta w_i^{(l)})+\overline{w}_i+\alpha(\overline{w}_i+\Delta w_i^{(r)})],\ 0\leq\alpha\leq 1,\ i\in\Omega \tag{7}$$

where the value α is an index of rating attitude. It reflects the decision maker's risk-bearing attitude. If $\alpha>0.5$, then it implies that decision maker is a risk lover. If $\alpha=0.5$, then the attitude of the decision maker is neutral to the risk. If $\alpha<0.5$, then the decision maker is a risk avertor. In general, the index of rating attitude, α, can be given by the decision maker directly.

Step 6. Rank all the alternatives x_i ($i=1,2,...,n$) and select the best one(s) in accordance with $\hat{w}_i^{(\alpha)}$ ($i\in\Omega$).

Step 7. End.

Acknowledgements

The author wishes to thank the anonymous referees for their careful reading of the manuscript and their fruitful comments and suggestions. This work was supported by China Postdoctoral Science Foundation (2003034366).

References

[0] T.L. Saaty: *The Analytic Hierarchy Process*. McGraw-Hill, New York, 1980.

[0] C. L. Hwang and K. Yoon: *Multiple Attribute Decision Making*. Springer-Verlag, Berlin, 1981.

[0] M. Zeeny: *Multiple Criteria Decision Making*. McGraw-Hill, New York, 1982.

[0] R.L. Keeney and H. Raiffa: *Decisions with Multiple Objectives*. Cambridge University Press, Cambridge, U.K, 1993.

[0] R.R.Yager and J. Kacprzyk: *The Ordered Weighted Averaging Operators: Theory and Applications*. Kluwer, Norwell, MA, 1997.

[0] Z.S. Xu: *Uncertain Multiple Attribute Decision Making: Methods and Applications*. Tsinghua University Press, Beijing, 2004.

[0] S. A. Orlovsky, Decision-making with a fuzzy preference relation, *Fuzzy Sets and Systems* **1** 155-167(1978).

[0] H. Nurmi, Approaches to collective decision making with fuzzy preference relations, *Fuzzy Sets and Systems* **6** 249-259 (1981).

[0] T. Tanino, Fuzzy preference orderings in group decision making, *Fuzzy Sets and Systems* **12** 117-131 (1984).

[0] J. Kacprzyk, Group decision making with a fuzzy linguistic majority, *Fuzzy Sets and Systems* **18** 105-118(1986).

[0] Z.S. Xu. Algorithm for priority of fuzzy complementary judgement matrix, *Journal of Systems Engineering* **16**(4) 311-314 (2001).

[0] F. Chiclana, F. Herrera and E. Herrera-Viedma, Integrating multiplicative preference relations in a multipurpose decision making model based on fuzzy preference relations, *Fuzzy Sets and Systems* **112** 277-291(2001).

[0] Z.S. Xu and Q.L. Da, The uncertain OWA operator, *International Journal of Intelligent Systems* **17** 569-575(2002).

[0] S. Lipovetsky and M. Michael Conklin, Robust estimation of priorities in the AHP, *European Journal of Operational Research* **137** 110-122(2002).

[0] Z.S. Xu and Q.L. Da, An approach to improving consistency of fuzzy preference matrix, *Fuzzy Optimization and Decision Making* **2** 3-12 (2003).

[0] E. Herrera-Viedma, F. Herrera, F. Chiclana and M. Luque. Some issues on consistency of fuzzy preference relations, *European Journal of Operational Research* **154** 98-109 (2004).

[0] Z.S. Xu, Goal programming models for obtaining the priority vector of incomplete fuzzy preference relation, *International Journal of Approximate Reasoning* **36** 261-270 (2004).

[0] Z.S. Xu and Q.L. Da, A least deviation method to obtain a priority vector of a fuzzy preference relation, *European Journal of Operational Research* **164** 206-216 (2005).

[0] Z.S. Xu, On compatibility of interval fuzzy preference matrices, *Fuzzy Optimization and Decision Making* **3** 217-225(2004).

[0] K. Yoon, The propagation of errors in Multiple-attribute decision analysis: a practical approach, *Journal of the Operational Research Society* **40** 681-686(1989).

[0] E.M. Pugh and G.H. Winslow: *The Analysis of Physical Measurements*. Addison-Wesley, Reading, MA, 1966.

[1] T.S. Liou, M. J. J. Wang, Ranking fuzzy numbers with integral value, *Fuzzy Sets and Systems* **50** 247-255(1992).

Brill Academic Publishers
P.O. Box 9000, 2300 PA Leiden,
The Netherlands

Lecture Series on Computer and Computational Sciences
Volume 2, 2005, pp. 150-154

Maximizing Deviations Procedure for Multiple Attribute Group Decision Making under Linguistic Environment

Z. S. Xu[1]

College of Economics and Management, Southeast University,
Nanjing, Jiangsu 210096, China

Received 8 March, 2005; accepted in revised form 16 March, 2005

Abstract: In this paper, we investigate the multiple attribute group decision making problems under linguistic environment, in which the information about attribute weights is unknown completely, and the attribute values (elements in each individual decision matrix) are in the form of linguistic variables. We first utilize the linguistic weighted averaging (LWA) operator to aggregate each individual linguistic decision matrix into a collective linguistic decision matrix. Then, we establish a single-objective optimization model maximizing deviations of the collective attribute values of all alternatives. By solving the model, a simple formula for determining attribute weights is obtained, and a practical procedure for ranking alternatives is also developed.

Keywords: Multiple attribute group decision making; linguistic variables; maximizing deviations.
Mathematics Subject Classification: 90D35, 65F15.

1. Introduction

Multiple attribute group decision making generally involves aggregate each individual decision information into the collective decision information and then obtains the most desirable alternative(s) by using the aggregated information. It is becoming more frequent and important in real world. In the process of group decision making, the decision makers are constantly providing linguistic preferences rather than numerical ones because of time pressure, lack of knowledge, and their limited attention and information processing capabilities [1-14]. Traditional approaches aggregating linguistic information have an important draw-the loss of information caused by the need to express the results in the initial discrete expression domain [10]. Recently, Herrera and Martínez [10-12] established a linguistic representation model for representing the linguistic information with a pair of values called 2-tuple, composed by a linguistic term and a number. Together with the model, they also presented a computational technique to deal with the 2-tuples without loss of information. Xu [15] developed an approach, which compute with words directly. Both these methods, however, are unsuitable for dealing with the problems in which the information about the attribute weights is unknown completely, and the attribute values take the form of linguistic variables. To overcome this drawback, we shall propose a practical procedure maximizing deviations for ranking alternatives.

[1] Corresponding author. E-mail: xu_zeshui@263.net

2. Procedure

Consider a finite and totally ordered discrete label set $S = \{s_\alpha \mid \alpha = -t, \ldots, t\}$, where s_α represents a possible value for a linguistic variable, and satisfies $s_\alpha > s_\beta$ if $\alpha > \beta$. For example, a set of nine labels S could be [16]:

$$S = \{s_{-4} = \textit{extremely poor},\ s_{-3} = \textit{very poor},\ s_{-2} = \textit{poor},\ s_{-1} = \textit{slightly poor},\ s_0 = \textit{fair},$$
$$s_1 = \textit{slightly good},\ s_2 = \textit{good},\ s_3 = \textit{very good},\ s_4 = \textit{extremely good}\}$$

The cardinality of S must be small enough so as not impose useless precision on the experts and it must be rich enough in order to allow a discrimination of the performances of each alternative in a limited number of grades [6].

In the process of information aggregating, some results may do not exactly match any linguistic labels in S. To preserve all the given information, Xu [14] extended the discrete label set S to a continuous label set $\bar{S} = \{s_\alpha \mid \alpha \in [-q, q]\}$, where q $(q > t)$ is a sufficiently large positive integer. If $s_\alpha \in S$, then s_α is termed an original linguistic label, otherwise, s_α is termed a virtual linguistic label. In general, the decision makers use the original linguistic labels to evaluate alternatives, and the virtual linguistic labels can only appear in operation.

Consider any two linguistic labels $s_\alpha, s_\beta \in \bar{S}$, and $\mu \in [0,1]$, we introduce two operational laws as follows[16]: 1) $\mu s_\alpha = s_{\mu\alpha}$; 2) $s_\alpha \oplus s_\beta = s_\beta \oplus s_\alpha = s_{\alpha+\beta}$.

To aggregate the given linguistic information, we shall introduce a useful aggregation operator called linguistic weighted averaging (LWA) operator as follows:

A linguistic weighted averaging (LWA) operator of dimension n is a mapping $LWA : \bar{S}^n \to \bar{S}$ such that

$$LWA_\omega(s_{\alpha_1}, s_{\alpha_2}, \ldots, s_{\alpha_n}) = \omega_1 s_{\alpha_1} \oplus \omega_2 s_{\alpha_2} \oplus \cdots \oplus \omega_n s_{\alpha_n} = s_{\bar{\alpha}} \tag{1}$$

where $\bar{\alpha} = \sum_{j=1}^{n} \omega_j \alpha_j$, $\omega = (\omega_1, \omega_2, \ldots, \omega_n)^T$ is the weighting vector of the $s_{\alpha_j} \in \bar{S}$, and $\omega_j \geq 0$, $\sum_{j=1}^{n} \omega_j = 1$.

In addition, we shall define the concept of distance between two linguistic variables:

Definition 1. *Let s_α and s_β be two linguistic variables, then we call*

$$d(s_\alpha, s_\beta) = |\alpha - \beta| \tag{2}$$

the distance between s_α and s_β.

The multiple attribute group decision making problem which is considered in this paper can be represented as follows:

Let $X = \{x_1, x_2, \ldots, x_n\}$ be a finite set of alternatives, $D = \{d_1, d_2, \ldots, d_l\}$ be the set of decision makers, and $\lambda = (\lambda_1, \lambda_2, \ldots, \lambda_l)^T$ be the weight vector of decision makers, where $\lambda_k \geq 0, \sum_{k=1}^{l} \lambda_k = 1$. Let $G = \{G_1, G_2, \ldots, G_m\}$ be a finite set of attributes, and $\omega = (\omega_1, \omega_2, \ldots, \omega_m)^T$ be the weight vector of attributes, where $\omega_i \geq 0, \sum_{i=1}^{m} \omega_i^2 = 1$. The information about attribute weights is unknown completely. Let $A^{(k)} = (a_{ij}^{(k)})_{m \times n}$ be the linguistic decision matrix, where $a_{ij}^{(k)} \in S$ is provided by the decision maker $d_k \in D$ for the alternative $x_j \in X$ with respect to the attribute $G_i \in G$.

To obtain the collective linguistic decision information, we utilize (1) to aggregate all individual linguistic decision matrices $A^{(k)} = \left(a_{ij}^{(k)}\right)_{m \times n}$ $(k = 1,2,...,l)$ into a collective linguistic decision matrix $A = \left(a_{ij}\right)_{m \times n}$, where

$a_{ij} = LWA_\lambda\left(a_{ij}^{(1)}, a_{ij}^{(2)}, ..., a_{ij}^{(l)}\right) = \lambda_1 a_{ij}^{(1)} \oplus \lambda_2 a_{ij}^{(2)} \oplus \cdots \oplus \lambda_l a_{ij}^{(l)}$ $i = 1,2,...,m;\ j = 1,2,...,n$

Based on the collective linguistic decision matrix $A = \left(a_{ij}\right)_{m \times n}$, the overall value of the alternative x_j can be expressed as

$$z_j(\omega) = \omega_1 a_{1j} \oplus \omega_2 a_{2j} \oplus \cdots \oplus \omega_m a_{mj}, \quad j = 1,2,...,n \tag{3}$$

Obviously, the greater the value $z_j(\omega)$, the better the alternative x_j will be.

In the situation where the information about attribute weights is completely known, we can rank all the alternatives by (3). However, in this paper, the information about the attribute weights in the problem considered is completely unknown. Thus, we need to determine the attribute weights in advance.

According to the information theory, if all alternatives have similar attribute values with respect to an attribute, then a small weight should be assigned to the attribute, because such attribute does not help in differentiating alternatives [17]. As a result, by (2), we introduce the deviation between the alternative x_j and all the other alternatives with respect to the attribute G_i:

$$d_{ij}(\omega) = \sum_{k \neq j} d\left(a_{ij}, a_{ik}\right)\omega_i, \quad i = 1,2,...,m;\ j = 1,2,...,n \tag{4}$$

Let

$$d_i(\omega) = \sum_{j=1}^{n} \sum_{k \neq j} d\left(a_{ij}, a_{ik}\right)\omega_i, \quad i = 1,2,...,m \tag{5}$$

denote the sum of all the deviations $d_{ij}(\omega)\,(j = 1,2,...,n)$, and construct the deviation function:

$$d(\omega) = \sum_{i=1}^{m} d_i(\omega) = \sum_{i=1}^{m} \sum_{j=1}^{n} d_{ij}(\omega) = \sum_{i=1}^{m} \sum_{j=1}^{n} \sum_{k \neq j} d\left(a_{ij}, a_{ik}\right)\omega_i \tag{6}$$

Obviously, a reasonable vector of attribute weights $\omega = \left(\omega_1, \omega_2, ..., \omega_m\right)^T$ should be determined so as to minimize $d(\omega)$, and thus, we establish the following single- objective optimization model:

$$\text{Maximize } d(\omega) = \sum_{i=1}^{m} \sum_{j=1}^{n} \sum_{k \neq j} d\left(a_{ij}, a_{ik}\right)\omega_i$$

$$\text{Subject to: } \sum_{i=1}^{m} \omega_i^2 = 1, \quad \omega_i \geq 0, \quad i = 1,2,...,m$$

To solve this model, we construct the Lagrange function:

$$L(\omega,\lambda) = d(\omega) + \frac{1}{2}\lambda\left(\sum_{i=1}^{m}\omega_i^2 - 1\right) \tag{7}$$

where λ is the Lagrange multiplier.

Differentiating (7) with respect to $\omega_i\ (i = 1,2,\ldots,m)$ and λ, and setting these partial derivatives equal to zero, the following set of equations is obtained:

$$\begin{cases} \dfrac{\partial L(\omega,\lambda)}{\partial \omega_i} = \sum_{j=1}^{n}\sum_{k\neq j} d(a_{ij}, a_{ik}) + \lambda w_i = 0 \\ \dfrac{\partial L(\omega,\lambda)}{\partial \lambda} = \sum_{i=1}^{m}\omega_i^2 - 1 = 0 \end{cases} \tag{8}$$

By solving (8), we get the optimal solution $\omega^* = (\omega_1^*, \omega_2^*, \ldots, \omega_m^*)^T$, where

$$\omega_i^* = \frac{\sum_{j=1}^{n}\sum_{k\neq j} d(a_{ij}, a_{ik})}{\sqrt{\sum_{i=1}^{m}\left(\sum_{j=1}^{n}\sum_{k\neq j} d(a_{ij}, a_{ik})\right)^2}}, \quad i = 1,2,\ldots,m \tag{9}$$

Obviously, $\omega_i^* \geq 0$, for all i. Normalizing (9), we get the normalized attribute weights:

$$\omega_i = \frac{\sum_{j=1}^{n}\sum_{k\neq j} d(a_{ij}, a_{ik})}{\sum_{i=1}^{n}\sum_{j=1}^{n}\sum_{k\neq j} d(a_{ij}, a_{ik})}, \quad i = 1,2,\ldots,m \tag{10}$$

By (3) and (10), we obtain the overall values $z_j(\omega)\ (j = 1,2,\ldots,n)$ of alternatives. Then, we rank all the alternatives by using $z_j(\omega)\ (j = 1,2,\ldots,n)$, and thus get the most desirable one(s).

Acknowledgements

The author wishes to thank the anonymous referees for their careful reading of the manuscript and their fruitful comments and suggestions. This work was supported by China Postdoctoral Science Foundation (2003034366).

References

[1] L.A. Zadeh. The concept of a linguistic variable and its application to approximate reasoning, *Part 1,2 and 3, Information Sciences* **8** 199-249, 301-357(1975); **9** 43-80(1976).

[2] J. Kacprzyk, Group decision making with a fuzzy linguistic majority, *Fuzzy Sets and Systems* **18** 105-118(1986).

[3] R. Degani and G. Bortolan, The problem of linguistic approximation in clinical decision making, *International Journal of Approximate Reasoning* **2** 143-162(1988).

[4] M. Delgado, J.L. Verdegay and M.A. Vila, On aggregation operations of linguistic labels, *International Journal of Intelligent Systems* **8** 351-370(1993).

[5] V. Torra, Negation functions based semantics for ordered linguistic labels, *International Journal of Intelligent Systems* **11** 975-988(1996).

[6] G. Bordogna, M. Fedrizzi and G. Passi, A linguistic modeling of consensus in group decision making based on OWA operator, *IEEE Transactions on Systems, Man, and Cybernetics* **27** 126-132(1997).

[7] L.A. Zadeh, J. Kacprzyk: *Computing with words in information/intelligent systems-part 1: foundations: part 2: applications*. Physica-Verlag, Heidelberg, Germany, 1999, vol. I.

[8] F. Herrera and E. Herrera-Viedma, Linguistic decision analysis: Steps for solving decision problems under linguistic information, *Fuzzy Sets and Systems* **115** 67-82 (2000).

[9] F. Herrera, E. Herrera-Viedma and L. Martínez, A fusion approach for managing multi-granularity linguistic term sets in decision making, *Fuzzy Sets and Systems* **114** 43-58(2000).

[10] F. Herrera and L. Martínez, A 2-tuple fuzzy linguistic representation model for computing with words, *IEEE Transactions on fuzzy systems* **8** 746-752(2000).

[11] F. Herrera and L. Martínez, A model based on linguistic 2-tuples for dealing with multigranular hierarchical linguistic contexts in multi-expert decision-making, *IEEE Transactions on Systems, Man, and Cybernetics* **31** 227-234(2001).

[12] M. Delgado, F. Herrera, E. Herrera-Viedma, M.J. Martin-Bautista, L. Martinez and M.A. Vila, A communication model based on the 2-tuple fuzzy linguistic representation for a distributed intelligent agent system on internet, *Soft Computing* **6** 320-328(2002).

[13] Z.S. Xu and Q.L. Da, An overview of operators for aggregating information, *International Journal of Intelligent Systems* **18** 953-969(2003).

[14] Z.S. Xu: *Uncertain Multiple Attribute Decision Making: Methods and Applications*. Tsinghua University Press, Beijing, 2004.

[15] Z.S. Xu, A method based on linguistic aggregation operators for group decision making with linguistic preference relations, *Information Sciences* **166** 19-30(2004).

[16] Z.S. Xu, Deviation measures of linguistic preference relations in group decision making, *Omega* **33** 249-254(2005).

[17] M. Zeleny: *Multiple Criteria Decision Making*. McGraw-Hill, New York, 1982.

[18] N. Bryson and A. Mobolurin, An action learning evaluation procedure for multiple criteria decision making problems, *European Journal of Operational Research* **96** 379-386(1995).

Brill Academic Publishers
P.O. Box 9000, 2300 PA Leiden,
The Netherlands

Lecture Series on Computer and Computational Sciences
Volume 2, 2005, pp. 155-158

Multicast Routing Path Selection in Fiber-optic Ring Networks: A Simulation Study

Rong-Jou Yang[1], Hann-Jang Ho

Department of Information Management
Faculty of Business
WuFeng Institute of Technology
117 Chienkao 2nd Rd.
Minhhsiung, Taiwan

Jack Lee

Dept. of Tech. Support and Training
Faculty of Business and Information Technology
Indian University of Pennsylvania
1011 South Drive
Indiana, USA

Received 11 April, 2005; accepted in revised form 18 April, 2005

Abstract: The multicast technology has emerged as a dominant and critical technique for the e-commerce applications to best utilize effectively the network bandwidth. In this paper, we propose a solution to the problem of multicast transmission in the bidirectional fiber-optic and non-fiber optic ring networks and study the performance improvement of multicast transmission in existing LAN/MAN. Given a set of multicast groups, our objective is to design a bandwidth allocation strategy for packet transmission that has the minimum traffic load. We formulate the problem as an ILP (Integer Linear Program) and proposed a heuristic algorithm to obtain certain approximate solution. The numerical results indicate that the algorithm provides generally certain of near-optimal solution comparing to other optimal solutions solved by ILP.

Keywords: Multicast Routing, Heuristic Rule, AMPL Simulation, Integer Linear Programming.

1. Non-Fiber Optic Ring Network

1.1 Definition of Symbols

Φ: The maximum flow of the network
n: The number of all the nodes
m: The number of all the multicast groups
hi: The ith multicast group
|hi|: The number of the member of the ith multicast group
Wi: The flow of the ith multicast group
Pij: The jth path used by the ith multicast group hi
YPij: The logical value indicates if the jth path used by the ith multicast group (1—Yes, 0—No)
Eh: The set of all the multicast groups {h1, h2, …, hm}
e: A connection between two adjacent nodes
P(e): The set of all the paths that pass connection e
Ec : The set of all the connections {E1, E2, …, En}

[1] Faculty of the Department of Information Management. E-mail: rjyang@mail.wfc.edu.tw

1.2 Constraints Formula

Using ILP to minimize Φ subjecting to:

Connectivity constraints:

$$\sum_{\substack{1 \le j \le |hi| \\ i \in \{1,2,\ldots,m\}}} yp_j^i \ge |hi| - 1, \qquad (1)$$

Capacity constraints:

$$\sum_{p_i^j \in P(e)} y_{p_i^j} w_i \le \Phi \ , \forall e \in \ Ec \qquad (2)$$

Binary variable: integer

$$0 \le y_{p_i^j} \le 1 \ , \forall i \in \ \{1,2,\ldots,m\}, j \in \ \{1,2,\ldots,|hi|\} \quad (3)$$

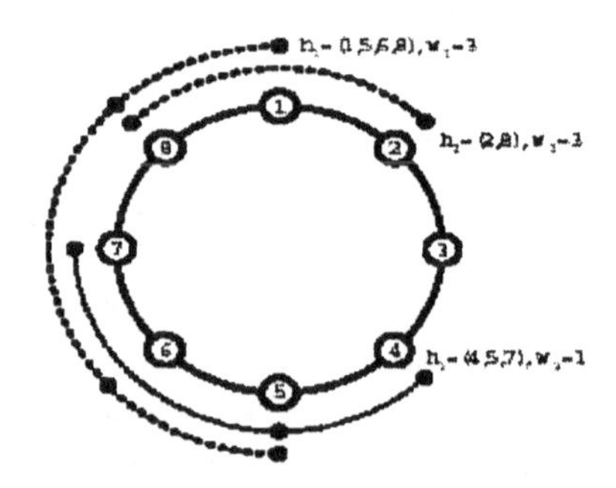

Figure 1: Example Non-fiber Optic Ring Network

1.3 The Resolution Steps by AMPL

Based on this heuristic rule, we can attain the optimal multicast routing path for the maximum flow of the ring network is: 6.

2. Fiber-optic Ring Network

2.1 Symbols Definition

n: The number of nodes

m: The number of multicast groups

hi: The ith multicast group

Eh: The set of all the multicast groups {h1, h2, …, hm}

x^w : If the light wave w is assigned (1: Yes, 0: No)

x_i^w : If the light wave w is assigned by the ith multicast group (1: Yes, 0: No)

$y_{p(i,j)}^w$: If the light wave w is assigned by the jth connection of the ith multicast group (1: Yes, 0: No)

Minimum Cost: The minimum number of light waves

2.2 Constraints Formula

$$1. \sum_j y_{p(i,j)}^w \ge (mi - 1)x_i^w \ , \forall hi \in Eh, w \in W \qquad (4)$$

The constraint of a light wave is continuous.

$$2. \sum_W x_i^w \ge r_i \qquad \forall hi \in Eh \qquad (5)$$

All the number of light waves is greater or equal to the required number of light waves. Otherwise, the solution will be intractable.

3. $$\sum_{p(i,j)\in p(e)} y^{w}_{p(i,j)} \leq x^{w} \qquad \forall w \in W, e \in E, x^{w} \in \{0,1\} \qquad (6)$$

There are only w light waves at most assigned for a connection.

4. $$x_{i}^{w} \leq x^{w} \qquad \forall w \in W, hi \in Eh \qquad (7)$$

A light wave is assigned for a connection only at a time.

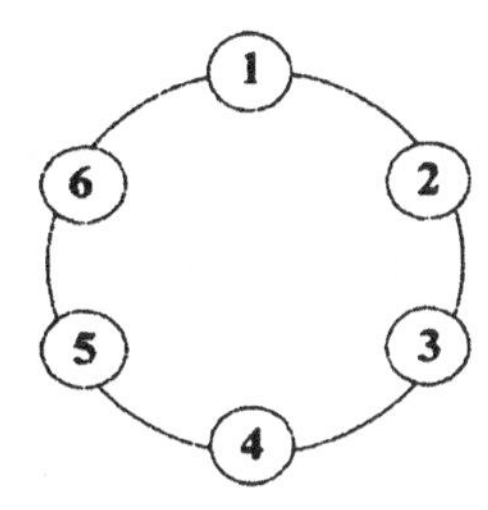

Figure 2: Example Fiber-optic Ring Network

2.3 The Comparison of Optimal Solution

In accordance with fiber-optic constraint equations and based on Figure 2, we also give another four example cases with non-integer of X or y to compare the differences listed as follows and the results of the minimum required light waves are indicated as in Table 1:

Case 1: 5 nodes, 4 multicast groups, and 8 light waves;
h1 (1, 2, 3) with 2 required light waves; h2 (2, 5) with 1 required light wave;
h3 (1, 3, 5) with 2 required light waves; h4 (2, 4, 5) with 3 required light waves;

Case 2: 5 nodes, 3 multicast groups, and 8 light waves;
h1 (1, 2, 3) with 2 required light waves; h2 (2, 5) with 4 required light wave;
h3 (1, 3, 5) with 2 required light waves;

Case 3: 6 nodes, 5 multicast groups, and 10 light waves;
h1 (1, 2) with 2 required light waves; h2 (2, 4, 6) with 2 required light waves; h3 (2, 3, 5) with 1 required light wave; h4 (1, 3, 6) with 1 required light wave; h5 (2, 5, 6) with 2 required light waves;

Case 4: 6 nodes, 4 multicast groups, and 8 light waves;
h1 (1, 2) with 2 required light waves; h2 (2, 4, 6) with 2 required light waves;

h3 (2, 3, 5) with 2 required light waves; h4 (1, 3, 6) with 2 required light waves;

Condition 1: Both X and y is integer.

Condition 2: Both X and y is floating point number.

Condition 3: X is floating point number and y is integer.

Condition 4: X is integer and y is floating point number.

Table 1: Minimum required light waves

Case / Condition	Case 1	Case 2	Case 3	Case 4
Condition 1	5	6	5	6
Condition 2	4	4	3.67	4
Condition 3	4	4	4	4
Condition 4	5	6	5	6

3. Conclusions and Future Extensions

In this paper, we probe into the NP-complete problem of multicast flow optimization in both the ring and fiber-optic ring network and try to solve the problem using AMPL instead of stressing on the comparison of the advantages and shortages of the different solutions as in traditional research method. We also propose a multicast routing algorithm in both the non-fiber optic and fiber-optic ring network to find the optimal point-to-point and multicast routing path with the minimization of the flow by using AMPL. In future extensions, two directions concerned on multicast routing path are still worth to be studied further and are as follows:

(1). Verify if the approximation solution resolved by AMPL is convergent to the optimal solution using mathematical method.

(2). Revise the multicast routing problem of minimization of maximum flow to support the dynamic joining and leaving the multicast group and find its optimal multicast routing path.

Acknowledgments

The authors wish to thank the anonymous referees for their careful reading of the manuscript and their fruitful comments and suggestions.

References

[1] Santosh Vempala and Berthold Vocking (1999), Approximating Multicast Congestion, *Proc. 10th Int. Symp, Algorithms and Computation,* pp. 367-372

[2] K. Jansen and H. Zhang (2002), An Approximation Algorithm for the Multicast Congestion Problem via Minimum Steiner Trees, *Proc. 3rd Int. Workshop Approximation and Randomized Algorithms in Communication Networks*

[3] S. L. Lee and H. J. Ho, On Minimizing the Maximum Congestion for Weighted Hypergraph Embedding in a Cycle, *Information Processing Letters,* 87(5), pp. 217-275 (2003)

[4] J. L. Ganley and J. P. Cohoon (1997), Minimum Congestion Hypergraph Embedding in a Cycle, *IEEE Trans. Comput.,* 46 (5): 600-602

[5] S. Cosares and I. Saniee (1992), An Optimization Problem Related to Balancing Loads on SONET Rings, *Telecommunications Systems, vol.3,* pp.165-181

[6] S. Khanna (1997), A Polynimial Time Approximation Scheme for SONET Ring Loading Problem, *Bell Labs Technical Journal,* Spring Issue, pp. 36-41

[7] L. A. Wolsey (1998), *Integer Programming,* Wiley.

[8] ILOG Optimization Suite White Paper, http://www.ilog.com/products/optimization/tech/optimization_whitepaper.pdf

[9] R. Fourer, D. M. Gay, and B. W. Kernighan (2002), *AMPL: A Modeling Language for Mathematical Programming,* Duxbury Press/Press/Cole Publishing Company, 2002

[10] T. Pusateri (1993), IP Multicast over Token Ring Local Area Networks, *Networks Working Group, RFC 1469(RFC/STD/FYI/BCP Archives)*

[11] Kanaiya Vasani (2001), Resilient Packet Rings, Foundation for a new MAN, *Computer Technology Review*

[12] Stenphen A Cook (1971), the Complexity of Theorem Proving Procedures. *Proceedings Third Annual ACM Symposium on Thoery of Computing,* pp 151-158

[13] Eric W. Weisstein, *CRC Concise Encyclopedia of Mathematics, Second Edition,* Chapman & Hall/CRC, December, 2002

Brill Academic Publishers
P.O. Box 9000, 2300 PA Leiden,
The Netherlands

Lecture Series on Computer and Computational Sciences
Volume 2, 2005, pp. 159-162

A Table Look-Up Pipelined Multiplication - Accumulation Co-Processor for H.264

Chang N. Zhang[1] and Xiang Xiao

Department of Computer Science
University of Regina
Telecommunication Research Labs (TRlabs)
Regina, SK, S4S0A2

Abstract: A newly developed video standard H.264/AVC provides significant better compression bit ratio and transmission efficiency [1][2]. To achieve these, much higher computation power is needed. The hybrid CPU/FPGA chip model contains CPU (DSP) and FPGAs can be used to reduce design cost and still support significant hardware customization. A H.264 Codec is implemented by a high end CPU (RM9000) and two FPGA implemented co-processors. This article presents a Table Look-Up Pipelined Multiplication-Accumulation Co-processor (MAC) for H.264 Codec, which can be used for a large amount of computations in H.264 with one of the multiplicands is a constant. The proposed design is based on the table looking up to generate the partial products and a three-level pipelined architecture is presented. A high-speed implementation of MAC within Xilinx FPGA [4] is explored. Performance analysis shows that the proposed FPGA implemented MAC can achieve output rates beyond 1.53 Gbit/s.

Keywords: H.264, video codec, table look-up, pipeline, FPGA, Verilog, System on Chip

1. Introduction

The early successes in the digital video industry (notably broadcast digital television and DVD-Video) were underpinned by international standard ISO/IEC 13818 [1], popularly known as MPEG-2'. Anticipation of a need for better compression tools has led to the development of two further standards for video compression known as ISO/IEC 14496 Part 2 (MPEG-4 Visual') [2] and ITU-T Recommendation H.264/ISO/IEC 14496 Part 10 (H.264') [3]. H.264/AVC is the newest video coding standard of the ITU-T Video Coding Experts Group and the ISO/IEC Moving Picture Experts Group. A H.264 Codec can be built by different hardware and software combinations. In our research, a H.264 Codec is built by a high end CPU (RM9000) and two FPGA implemented co-processors. They are: (1) Pipelined Variable-Window Size Motion Estimation Co-processor and (2) Table Look-Up Pipelined Multiplication-Accumulation Co-processor, which is to be presented in this paper. According to our studies, a large amount of computations in H.264 can be fully or partially converted into a uniform integer multiplication-accumulation formula with one of the multiplicators is a constant. The unique features of the Table Look-Up Pipelined Multiplication - Accumulation Co-processor include (1) Each product is decomposed into two same length partial products which are generated by two precomputed look-up tables to reduce the circuit complexity and improve the performance. (2) All partial products are summed by a 4-3 compressor and carry save adder array to eliminate the carry propagations. (3) A three-stage pipeline architecture is used to maximize the throughputs. A FPGA implementation of the MAC co-processor is presented in section 3. A Xilinx FPGA development board is used with on-board SDRAM as the look-up tables and the software CPU as the Co-processor

[1] Corresponding author. Professor of Department of Computer Science, University of Regina, Canada. E-mail: zhang@cs.uregina.ca.

controller. The performance shows that the proposed co-processor works well and generates up to 1.53Gbit/s at the clock of 192MHZ.

2. Circuit Design of MAC

We have designed a Multiplication - Accumulation Co-processor to implement the following sum of products calculation:

$$Y = \sum_{i=1}^{n} A_i X_i \tag{1}$$

Where A_i is a 12-bit constant integer and X_i is an 8-bit integer. Both are in 2's complement format. n = 4, 8, 16 or 32. Our goal is to design a high-performance circuit with acceptable cost.
Note that $X = X_7\ X_6\ X_5\ X_4\ X_3\ X_2\ X_1\ X_0 = -2^7 * X_7 + X_6 * 2^6 + X_5 * 2^5 + X_4 * 2^4 + X_3 * 2^3 + X_2 * 2^2 + X_1 * 2^1 + X_0 * 2^0$. Let $W = AX = Z + S * 2^4$ where

$$W = AX = -2^7 * A * X_7 + A * (X_6\ X_5\ X_4) * 2^4 + A * (X_3\ X_2\ X_1\ X_0) \tag{2}$$

1. If $X_7 = 0$ ($X \geq 0$), then $Z = A * (X_3\ X_2\ X_1\ X_0)$ and $S = A * (X_6\ X_5\ X_4)$.
2. If $X_7 = 1$ ($X < 0$), then $Z = -A * (\overline{X_3}\ \overline{X_2}\ \overline{X_1}\ \overline{X_0} + 1)$ and $S = -A * (\overline{X_6}\ \overline{X_5}\ \overline{X_4})$.

Since A is a constant, the partial products Z and S can be obtained from two precomputed Look-Up Tables (LUTs). Note that:

1) When n = 32 A has 32 possible values, we need 5 address bits (N_0, N_1, N_2, N_3 and N_4) to specify which constant A to be used in the calculation.

2) The other inputs to the two LUTs are $X_7\ X_6\ X_5\ X_4$ and $X_7\ X_3\ X_2\ X_1\ X_0$ respectively. With above parameters, we can precalculate the partial products and store them in the two LUTs.

2.1. Look-Up Tables

In our design, 5 bits of LUT inputs (N_0, N_1, N_2, N_3 and N_4) select which constant is used, while the remaining inputs are for the variable multiplicand. The two LUTs to generate partial products of Z and S are precomputed and loaded. The product A*X is 20-bit long and there will be maximum 32 products. So the final result Y(X) is 25-bit long. However we do not need store all 25 bits for the partial products Z and S since each partial product is only 16 or 15 bits long and the other higher bits are all '0' or '1'. So we can use only one extra bit to record the value of all the higher bits to save memory space.
In the first LUT (generating partial product Z), there are 5 inputs: $X_3\ X_2\ X_1\ X_0$ and X_7 (X_7 is used to determine whether X is positive or negative. If $X_7 = 1$, then X is negative and we should use $-(\overline{X_3}\ \overline{X_2}\ \overline{X_1}\ \overline{X_0} + 1)$ as the multiplier. If $X_7 = 0$, then we use $X_3\ X_2\ X_1\ X_0$ as the multiplier.) The outputs of the first LUT are 17 bits long ($Z_{16} \ldots Z_0$). Ai is addressed by the control signals N_0, N_1, N_2, N_3 and N_4. The size of the first LUT is $2^{10} \times 17 = 1024 \times 17$.
In the second LUT (generating partial product S), input signals are $X_7\ X_6\ X_5\ X_4$. The outputs of the second LUT are 16 bits long ($S_{15} \ldots S_0$). When $X_7 = 1$, X is negative and we should use $-(\overline{X_6}\ \overline{X_5}\ \overline{X_4})$ as the multiplier. When $X_7 = 0$, we use $X_6\ X_5\ X_4$ as the multiplier. Ai is addressed by control signals N_0, N_1, N_2, N_3 and N_4. The size of the second LUT is $2^9 \times 16 = 512 \times 16$.

2.2. Three-Level Pipelined Design

The pipelined architecture of MAC offers the advantage of extremely high throughput rates: once the latency of the pipeline has been met, the system will finish one accumulation of multiplication at each

clock cycle. The diagram of pipeline design for Table Look-Up Pipelined Multiplication - Accumulation Co-processor is shown in the Figure 1.
The pipeline is divided into 3 levels from top to the bottom and each level is to implement a specific function. Detailed functions of each level are described as follows.

1) The first level is to get the two partial products ($S_{15} \ldots S_0$) and ($Z_{16} \ldots Z_0$) by two LUTs.

2) The second level is to implement accumulation by adding the previous summation represented by carry save form of C_i and S_i (i = 0...24) with the present two partial products S and Z. The 4-3 Compressors and Carry Save Adders (CSA) are used to eliminate the carry propagations. Figure 4 shows the logic diagram of 4-3 compressor which has four inputs: I_0、I_1、I_2、I_3 and 3 outputs: D, C_0 and C_1. Inputs and outputs satisfy equation of $D + 2 \times C_0 + 2 \times C_1 = I_0 + I_1 + I_2 + I_3$.

3) The third level is to complete the calculation with a Carry Look Ahead Adder (CLA) and get the final result. We only need to calculate the final result after every n (n = 4, 8, 16 or 32) multiplications (A*X) instead of doing after each pipeline clock.

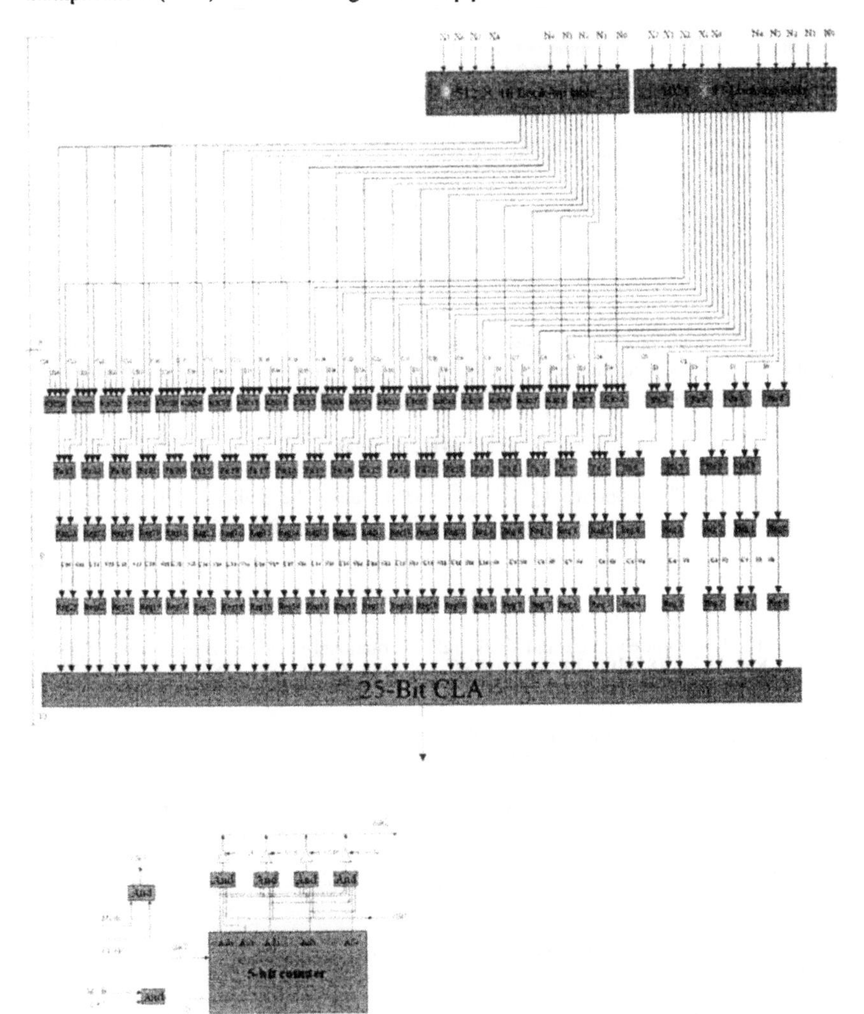

Figure 1. Diagram of Table Look-Up Pipelined Multiplication-Accumulation Co-processor

3. FPGA Implementation and Performance Analysis

The hardware impleme l ntation of the multiply - accumulation needs considerable resources. Several designs have been suggested. Scaling Accumulator Multiplier (SAM) is one of them. SAM is based on reducing the size of multiplier and performs the operations serially. This approach takes the first operand as a parallel load and others in serially. The result of this approach is shifted and accumulated with the old result. The Xilinx Virtex-II Multimedia development board was chosen as the target device. Resource utilization characteristics and timing characteristics must be considered when evaluating the performance of FPGA implementations of multiply-accumulate calculation. Table 1 details the resource utilization characteristics for each of the hardware implementation in number of CLB (Configurable Logic Block) slices.

Table 1. Resource utilization characteristics on the XC2V2000

Architecture	CLB slices	Utilization
MAC	87	8.25%
SAM	178	16.89%

Table 1 shows that the MAC implementation 1 is much smaller than the one for SAM. But MAC needs 2 extra RAM blocks on the FPGA board. Table 2 details the timing characteristics for the multiply-accumulate calculation for each of the hardware implementations.

Table 2. Timing characteristics on the XC2V2000

Architecture	Frequency	Throughput
MAC	192 MHZ	1.53 Gbit/s
SAM	86 MHZ	0.68 Gbit/s

From table 2, we get that MAC can operate at 192 MHZ clock (determined by the longer delay of level 1 and level 2 in the pipelined design). As the MAC is able to process one 4x4 pixel block of 8 bit values in 2 clock cycles, a maximum processing speed of 1.53 Gbit/s results on the target FPGA. From tables 1 and 2, it is of interest to note that MAC outperforms SAM in both area and speed.

Acknowledgments

This project is partially supported by NSERC and TRlabs. The authors also wish to thank the anonymous referees for their careful reading of the manuscript and their fruitful comments and suggestions.

References

[1] "Draft ITU-T recommendation and final draft international standard of joint video specification (ITU-T Rec. H.264/ISO/IEC 14 496-10 AVC," in Joint Video Team (JVT) of ISO/IEC MPEG and ITU-T VCEG, JVTG050, 2003.

[2] T. Stockhammer, M. M. Hannuksela, and T. Wiegand, "H.264/AVC in wireless environments," IEEE Trans. Circuits Syst. Video Technol., vol. 13, pp. 657–673, July 2003.

[3] T. Wiegand, and A. Luthra, Overview of the H.264 / AVC Video Coding Standard, IEEE transactions on Circuit and Systems for Video Technology, 2003.

[4] Goslin, G. R. &Newgard, B. "16-Tap, 8-Bit FIR Filter Application Guide," Xilinx Publications, 21 Nov94.

[5] ISO/IEC 13818, Information Technology- Generic Coding of Moving Pictures and Associated Audio Information, 2000.

[6] H. Malvar, A. Hallapuro, M. Karczewicz, and L. Kerofsky, "Low-Complexity transform and quantization in H.264/AVC," IEEE Trans. Circuits Syst. Video Technol., vol. 13, pp. 598–603, July 2003.

Brill Academic Publishers
P.O. Box 9000, 2300 PA Leiden,
The Netherlands

Lecture Series on Computer and Computational Sciences
Volume 2, 2005, pp. 163-165

Computers & Security

Guest Editor's Introduction to Special Session

Nicolas Sklavos [1]

Electrical & Computer Engineering Department,
University of Patras, Greece

1. Introduction: Computers & Security

Computers and networks attacks have become pervasive in today's communications world. Any personal computer or laptop connected to Internet is under the threat of hackers, eavesdroppers and attacks. Home users and especially business users, are attacked on a regular basis. It is obvious that has become an issue of high level importance, the need to combat computer and networks attacks.

Although traditionally cryptography has been associated with security and privacy in communications, it actually has many other objectives, among them such as authentication and data integrity. The first guarantees the sender's identity, while the second one guarantees that data have not been altered.

From the theoretical point of view, today cryptography can solve security problems. Although, the new applications scenarios for communications give rise to new demands. This Special Session is aimed to give solutions to today's security problems, via flexible and trustworthy solutions.

2. Special Session Call

This Special Session was organized as a part of the *International e-Conference of Computer Science 2005 (IeCCS 2005),* May 12-17, 2005. The related issues of the Special Session Call for the requested submitted manuscripts were:

- *Cryptography*
- *Modelling and Architecture*
- *Adaptive Security*
- *Public Key Infrastructure*
- *Cryptanalysis*
- *New encryption algorithms*
- *Case Studies, Surveys*
- *Security Schemes*
- *Implementation Cost and Performance Evaluation*

For all the above research topics, we received a great number of submissions. Finally, and due to space issues, only few papers were accepted in order to be presented in this Special Session.

The first accepted paper [3] of this Special Session is from M. Shahraki. This work focuses on the alternative integration approaches of AES suitable for software and hardware applications. It presents

[1] Corresponding author, email:nsklavos@ee.upatras.gr

and advantages and the trade offs, for alternative approaches, as well as comparisons results concerning performance issues. Furthermore this work introduces the design criteria for AES developments in both software and hardware manners.

A low power implementation of SHA-256 hash function is given in [4], in which the achieved throughput exceeds the limit of 2 Gbps. Furthermore, power dissipation is kept low in such way that the proposed implementation can be characterized as low-power.

The next paper [5] is related to Switchable Data-Dependent Operations and especially to the New Designs based on them.

A. Rjoub, et al., presents a high bit modular multiplier architecture in [6]. The design is based on the Radix-4 Montgomery multiplication algorithm that uses two types of digit recoding which results in a competitive design.

The work of N. A. Moldovyan [7] focuses the problem of increasing the integral VLSI implementation efficacy of block ciphers, which are suitable to applications where constrained resources are available. In this paper a new 64-bit cipher Eagle64 is developed, based on the approach of Data Dependent (DD) Operations (DDOs).

P. Kitsos in [8] proposes a reconfigurable Linear Feedback Shift Register (LFSR) architecture for Galois field $GF(2^m)$. The advantages of this LFSR are (i) the high order of flexibility, which allows an easy configuration for different field sizes, and (ii) the low hardware complexity, which results in small area.

3. Outlook

Last but not at least I would like to thank the Committee and especially the Chairman Prof. T. E. Simos of the *International e-Conference of Computer Science 2005 (IeCCS 2005)*, for hosting this Special Session.

In addition, special thanks to the reviewers of the submitted manuscripts. Their efforts helped all the authors to improve the quality of their works.

Finally many thanks to the authors. Their introduced ideas contributed to the research community at a great factor, in order to go a step forward to science, research, and engineering.

Dr. Nicolas Sklavos

Guest Editor

Dr. Nicolas Sklavos received the Ph.D. Degree in Electrical & Computer Engineering, and the Diploma in Electrical & Computer Engineering, in 2004 and in 2000 respectively, both from the Electrical & Computer Engineering Dept., University of Patras, Greece. His research interests include Cryptography, Wireless Communications Security, VLSI Design, and Reconfigurable Computing Architectures. He holds an award for his PhD thesis on "VLSI Designs of Wireless Communications Security Systems", from IFIP VLSI SOC 2003. He serves as a committee member of journals conferences and as referee also. He has organized several Special Sessions for international journals and conferences, in the areas of his research. Dr. N. Sklavos is a member of the IEEE, the Technical Chamber of Greece, and the Greek Electrical Engineering Society. He has authored or co-authored more than 80 scientific articles, book chapters, and reports, in the areas of his research. Contact him at Electrical & Computer Engineering Dept, University of Patras, Greece; *email: nsklavos@ee.upatras.gr*.

References

[1] B. Schneier, *Applied Cryptography–Protocols, Algorithms and Source Code in C*, Second Edition, John Wiley and Sons, New York, (1996).

[2] N. Sklavos, and O. Koufopavlou, Mobile Communications World: Security Implementations Aspects – A State of the Art, *CSJM, Institute of Mathematics and Computer Science*, 11, 32, 168-187, (2003).

[3] M. Shahraki, Implementation Aspects of Rijndael Encryption Algorithm, proceedings of *International e-Conference of Computer Science 2005 (IeCCS 2005)*.

[4] H. Michail, et al., Low Power and High Throughput Implementation of SHA-256 Hash Function, proceedings of *International e-Conference of Computer Science 2005 (IeCCS 2005)*.

[5] N.A. Moldovyan, et al., Switchable Data-Dependent Operations: New Designs, proceedings of *International e-Conference of Computer Science 2005 (IeCCS 2005)*.

[6] A. Rjoub, et al., A Low Power, High Frequency and High Precision Multiplier Architecture in GF(p), proceedings of *International e-Conference of Computer Science 2005 (IeCCS 2005)*.

[7] N. A. Moldovyan, et al., A New 64-bit Cipher for Efficient Implementation Using Constrained FPGA Resources, proceedings of *International e-Conference of Computer Science 2005 (IeCCS 2005)*.

[8] P. Kitsos, et al., A Reconfigurable Linear Feedback Shift Register Architecture for Cryptographic Applications, proceedings of *International e-Conference of Computer Science 2005 (IeCCS 2005)*.

Brill Academic Publishers
P.O. Box 9000, 2300 PA Leiden,
The Netherlands

Lecture Series on Computer and Computational Sciences
Volume 2, 2005, pp. 166-169

Implementation Aspects of Rijndael Encryption Algorithm

M. Shahraki

Department of Electrical Engineering,
University of Sistan & Baluchistan
Email: shahraki@rediffmail.com

Abstract: A cryptographic system is not only needed to protect an stand-alone system, like databases on a PC, operating system, system drivers and programs running on a system, but also it is needed to protect information transmitted between a set of devices, like; router, ATM switches etc [1]. Each one of these applications has unique characteristics and therefore need unique features that cryptography must satisfy. Advanced Encryption Standard (Rijndael) is examined in this paper, concerning software and hardware implementation platforms. This work focuses on the alternative integration approaches of Rijndael suitable for software and hardware applications. It presents the advantages and the trade offs, for alternative approaches, as well as comparison results concerning performance issues. Furthermore this work introduces the design criteria for AES developments in both software and hardware manners.

Keywords: Advanced Encryption Standard, AES, Software, Hardware, Networks, Communications

1. Introduction

Advanced Encryption Standard Ciphers consist of Rijndael, Serpent, Twofish, RC6 and Mars, which Rijndael is the winner of this group and is selected as the future algorithm to protect applications against adversaries. These applications range from wired to wireless [1] systems which all need encryption to transmit their information very secure and fast against adversaries and their attacks. Each one of these applications has unique characteristics and therefore need unique features that cryptography must satisfy. Advanced Encryption Standard (Rijndael) is examined in this paper, concerning software and hardware implementation platforms and their pros and cons.
First of all in section 2, its software implementation, concerning different processor platforms [2, 3, 4] and different soft tools (assembler/compiler) [5, 6] is presented. In section 3, its hardware implementation, concerning different methods, architecture [7] and platforms is noticed. In section 4 other criteria which are important in designing a protected system is considered. We will see that in some applications to satisfy required characteristics, it is needed to combine hardware and software design for a cryptographic purpose.

2. Software Implementation of Rijndael

Cryptography system, in this article, we mean Rijndael, could be implemented as a software program. Such implementation has public advantages, like; ease of use, ease of maintenance, portability and low cost, but it offers disadvantages like; poor security, higher power consumption and lower speed against hardware implementation [1]. Software implementation is influenced with language, compiler, platform, block size, key size, implementation structure and designer's used methodology. It could use C, Java, Assembly and Matlab as a soft tool [2, 3, 5, 6], and also a processor (SISD, SIMD and MIMD) [3], or MicroController (like 8051 [4]), as a hard platform.

Table 1: Clock cycles required for AES

Processor	Key schedule and Enc. / Dec. using key unrolling		Enc. / Dec. using key on-the-fly	Implemented in
ARM7TDMI	634	1675 / 2074	2074 / 2378	[6]
	449	1641 / 2763	1950 / 3221	[5]
ARM9TDMI	499	1384 / 1764	1755 / 1976	[6]
	333	1374 / 2439	1623 / 2796	[5]
ST22	0.22	0.51 / 0.60	0.72 / 0.82	[6]
	0.13	0.61 / 1	0.75 / 1.13	[5]
Pentium III	370	1119 / 1395	-	[6]
	396	1404 / 2152	-	[5]

Table 1, shows the effect of different hard platforms on encryption, decryption and key scheduling algorithms of Rijndael. By optimizing the algorithm it is possible to enhance software speed and memory usage [2], but however this enhancement is not considerable.
Hardware platform can be upon RISC or CISC. RISC has short and simple instructions and regular form in fetching/decoding, pipelining and scheduling, that enables a fast and simple implementation, but it is also fast and simple for adversaries to attack this implementation [8]. Basic parameters that influence software implementation with a special hard platform are: basic underlying architecture, whether it is 32 bits, 8 bits etc, use of different memories, structure of processor (SISD/SIMD/IMD), and its internal operation, number of Cpu ports and the number and length of registers.
As indicated, a soft tool could be Java, C, assembly, Matlab etc. In Java programming (32 bit processor), source code is compiled into byte code instead of machine code [2], so at runtime byte code must be converted to machine code, because Java programs are independent of processor and operating system. Therefore, a sizeable portion of compilation is delayed until runtime that this characteristic slows down Java compared to compilers like; C and also assemblers. Also Java supports 32 bits processors against C and assembly which support 64 bit processors, and this feature influences speed of Rijndael cryptography algorithms execution. On the other hand assembly code has better results (code size and speed) than that the C, because of optimized use of internal architecture of processor and use of complex instructions. Matlab is another alternative [5], because of its best representation for key and blocks of plain/cipher texts.
Anyway key scheduling of Rijndael is about 7 times faster than other AESs, its encryption/decryption is lower than RC6 but not more than 32%. In Java and C implementations, encryption/decryption algorithm of Rijndael is not the fastest, because Rijndael is not c-friendly, but in assembly, Rijndael makes very heavy use of the processors' new technologies. Rijndael is based only on most simple imaginable operations like; load and Xor. On the other hand Rijndael has large internal parallelism ability, and there is a large number of possibilities of reschedule its code, table 2 [6].

Table 2: C implementation of Rijndael compared to other AESs.

Cipher	Mbps on a 450 MHz PII	Clock/block	μ OPs/cycle
Rijndael	243	237	2.54
RC6	258	223	1.47
Twofish	204	282	2.11
Mars	188	306	1.87
Serpent	-	-	-

3. AES Hardware Implementation

Rijndael could be implemented in different modes and architectures, and in a variety of hardware devices like; ASICs, FPGA and CPLD, and also Smart Cards. Hardware platforms work with less frequency than current processors, but it is generally faster than their software equivalent, about 4 times [9], because Rijndael has parallel processing and pipeline characteristics that well suit on hardware platform, of course this is not common for all ciphers. Against to the tight physical security, low power consumption, high execution speed, some cons like difficult implementation, high implementation cost, and etc exist in hardware implementations. These factors are critical items, which must be taken care of special attention of the designers. ASIC designs guarantee better performance, with fast execution, enough small dedicated size and security against software implementation and even FPGA, but it is less feasible regard to both of them. Also FPGA provide faster and easier design, more flexibility and reconfigurability, which is very important for key-agility, as like we see on SSL, TLS etc [1]. FPGA is a middle one that has flexibility of software design and security and speed of hardware design. But even FPGA could be configured at runtime, decide number of pins is necessary before board implementation. Smart card is another option for hardware implementation of Rijndael, but in smart cards the RAM requirements are more important, than the clock frequency. It is cleared that the devices of this category are not proper for large encryption systems with special specifications [1] because of its slow communication with external memories.
Anyway, hardware implementation could be done in feedback (CBC, CFB, and OFB), and non-feedback modes (ECB), and architecture for the encryption/decryption unit could be implemented in one of these methods: Basic iterative architecture, Partial and full loop unrolling, Outer-round pipelining, Resource Sharing. Researches performed in [7] and similar efforts show that in iterative architecture and feedback modes, Serpent and Rijndael have the highest throughput at the expense of the relatively large area. On the other hand Rijndael and Mars use the modest area in feedback mode and iterative architecture [7], as indicated in table 3. Differences between different implementations of

Rijndael is according to different optimization for speed, area and sharing resources, which depends on designer's art, but no many improvement could be achieved.

Table 3: Implementations Comparisons

Cipher	throughput (Mbit/s)	Area (CLB slices)	Throughput/Area ration
RC6	142.7	1137	0.122
Twofish	177.3	1076	0.164
Rijndael	414.2	2507	0.166
Serpent I1	-	-	0.124
Serpent I8	431.4	4507	0.097
Mars	61.0	2744	0.022
3DES	59.1	356	-

For feedback mode, iterative architecture is very suitable, but decision to choose architecture for non-feedback modes is not easy [7], different architecture in non-feedback modes has different effects on Rijndael, table 4 shows affect of different architecture modes in non-feedback mode for Rijndael. In non-feedback mode, throughput is in excess of 3.65 Gbit/s [10], taking into account both FPGA and ASIC implementations. Full mixed-inner-outer round implementation has better throughput than other architecture. Implementation of Rijndael consumes approximately the same amount of FPGA resources like Serpent and Twofish. No correlation between software and hardware performance was found. The difference among Rijndael implementations in software or hardware is based on internal structure of these algorithms.

Table 4: Different hardware implementation of Rijndael

Architecture	Throughput (Mbits/s)
Iterative (feedback mode) [4]	414.2 Mbit/s
Full mixed -inner-outer -round pipelining (non-feedback mode) [4]	12.2 Gbit/s
Full outer-round pipelining (non-feedback mode) [4]	5.7 Gbit/s
Rijndael [10]	3.65 Gbit/s
Rijndael processor [11]	2.29 Gbit/s

4. Other Criteria in Rijndael Implementations

The performance of cryptography in high-speed applications closely requires tradeoff between security and speed, and there are many criteria that must be considered in software/hardware implementation to be able to have better performance and throughput. In some applications speed is more important than other features, and in some of them, power consumption and in a group of applications which have algorithm-independent nature, like; Secure Sockets Layer (SSL), IPsec [1], voice-over-IP products, high-speed routers and ATM switches [1, 7], switching between Rijndael and other ciphers is very important. Rijndael could be implemented to present high speed and low power consumption, e.g. it could present higher speed and lower power consumption and used area for 802.11 protocols, against RC4 [10]. Also Rijndael could be implemented with different hardware methodologies to satisfy low power consumption or high execution speed. In [11] two different VLSI architectures are presented. The first uses feedback logic and reaches throughput value equal to 259 Mbit/sec with low covered area resources and the second is optimized for high-speed performance using pipelining technique with high data throughput of 3.65 Gbit/sec. These two implementations could be used for online cryptographic needs of high speed networking protocols.

But anyway, against all consideration to speed up and decrease covered area, in before sections, there are many attacks from adversaries' side both for software [8, 12] and hardware [8, 13] implementations of not only Rijndael [14] but also all ciphers. Computer hackers often have many techniques, either in hardware and/or software, at their disposal to crack out the secret [8]. Software implementation of cryptography algorithms is based on an operating system, therefore it maybe execute in parallel with other processes that this subject slows it down, and cause to use a common memory (main memory for holding intermediate results and external memory to hold encryption/decryption modules and long term keys). So there is no protection for code, key, and also intermediate results. Power consumption analysis and reverse engineering are other attacks that could disturb software implementations of ciphers and other applications. Also there are some known and special attacks for Rijndael, like; Impossible Differential attack, Square attack and Collision attack [14].

Hardware implementation of Rijndael, does not have software weaknesses which are presented in last paragraph, but it is natural that similar to software side, there are some attacks against hardware implementations on Smart Cards, ASICs and FPGAs, See [8,13] for more details. Comparing attacks against hardware and software implementation concludes that FPGA and ASICs are currently more secure than its software implementation peers.

Another concept in Rijndael implementation is its co-design. It is very noticeable, because this type of design could be used to improve final product security and speed, especially when key agility between Rijndael and other ciphers is needed.

6. Conclusion

In step with high performance networks and applications, high speed cryptography is needed more than before. It must be considered that high performance applications require an optimal trade off between security and speed. Fast hardware relies on parallelism and pipelining, while in software designs access to memory is a key to gain fast performance. This aspect becomes more and more important as the access time to the memory seems to decrease more slowly than the cycle time of the processor. In addition, it is presented in this paper that there are many more criteria which must be consider when a designer wants to develop a cryptographic system for an application. It could be implemented in software, hardware or a co-design manner, with a variety of algorithms and platforms.

Acknowledgments

The author wishes to thank the anonymous referees for their careful reading of the manuscript and their fruitful comments and suggestions.

References

[1] N. Sklavos, PhD Thesis on *"VLSI Designs of Wireless Communications Security Systems"*, proceedings of 12th International Conference on Very Large Scale Integration, (IFIP VLSI SOC '03), Darmstadt, Germany, December 1-3, 2003.

[2] Eashwar Thiagarajan and Madhuri Gourishetty, *"Study of AES and its Efficient Software Implementation"*, Department of Electrical Engineering & Computer Science, Oregon State University, Corvallis, Oregon 97331 -USA, 2001.

[3] Guido Bertoni1, Luca Breveglieri1, Pasqualina Fragneto, Marco Macchetti, and Stefano Marchesin *"Efficient Software Implementation of AES on 32-Bit Platforms"*, in Cryptographic Hardware and Embedded Systems - CHES 2002, pp. 159-171, B.S. Kaliski Jr., .K. Ko, C. Paar.

[4] Chi-Feng Lu, Yan-Shun Kao, Hsia-Ling Chiang, Chung-Huang Yang, *"High Speed Software Driven AES Algorithm on IC Smartcards"*, SCIS 2004 The 2004 Symposium on Cryptography and Information Security Sendai, Japan, Jan.27-30, 2004

[5] Andreas Sterbenz, Peter Lipp, *"Performance of the AES Candidate Algorithms in Java"*, AES Candidate Conference 2000, pp. 161-165.

[6] Kazumaro Aoki, helger Lipmaa, *"Fast Implementation of AES Candidates"*, in submitted for publication-third AES Candidate Conference, New York City, USA, August 13-15, 2001.

[7] Kris Gaj and Pawel Chodowiec, *"Fast implementation and fair comparison of the final Candidates for Advanced Encryption Standard using Field Programmable Gate Arrays"*, Proc. RSA Security Conference - Cryptographer's Track San Francisco, CA, April 8-12, 2001.

[8] Weidong Shi, Hsien-Hsin S. Lee, Chenghuai Lu, and Mrinmoy Ghosh, *"Towards the Issues in Architectural Support for Protection of Software Execution"*, ACM SIGARCH Computer Architecture News, pp. 6-15, 2005, ISSN: 0163-5964.

[9] Piotr Mroczkowski, *"Implementation of the block cipher Rijndael using Altera FPGA"*, Military University of Technology, Warsaw, Poland, 2000, pmrocz@mamut.isi.wat.waw.pl

[10] N. Sklavos, G. Selimis and O. Koufopavlou, *"FPGA Implementation Cost & Performance Evaluation of IEEE 802.11 Protocol Encryption Security Schemes"*, Proceeding of Second Conference on Microelectronics, Microsystems and Nanotechnology, (MMN'04), November 14-17, Athens, Greece 2004.

[11] N.Sklavos, O. Koufopavlou, "Architectures and VLSI Implementations of the AES-Proposal Rijndael", IEEE Transactions On Computers, Vol. 51, No. 12, December 2002.

[12] Hagai Bar-El, *"Security Implecations of Hardware vs. Software Cryptographic Modules"*, http://www.discretix.com, October 2002.

[13] Thomas Wollinger, Christof Paar, *"How Secure are FPGAs in Cryptographic Applications?"* FPL 2003, Vol. 2778, pp.91-100, 2004.

[14] Elisabeth Oswald, Joan Daemen, Vincent Rijmen, *"AES - The State of the Art of Rijndael's Security"*, October 30, 2002.

Brill Academic Publishers
P.O. Box 9000, 2300 PA Leiden,
The Netherlands

Lecture Series on Computer and Computational Sciences
Volume 2, 2005, pp. 170-173

Low Power and High Throughput Implementation of SHA-256 Hash Function

H.E.MICHAIL, A.P.KAKAROUNTAS, A.MILIDONIS, C.E.GOUTIS[1]

Electrical & Computer Engineering Department
University of Patras,
GR-26500 Patra, Greece

Abstract: Hash functions are utilized in the security layer of every communication protocol and in signature authentication schemes for electronic transactions.As time passes more sophisticated applications arise that address to more users-clients and thus demand for higher throughput.Furthermore, due to the tendency of the market to minimize devices size and increase their autonomy to make them portable, power issues should also be taken into consideration. Long rumored and now official, the popular and widely used SHA-1 hashing algorithm has been attacked successfully by researchers in China and the US.It is obvious that sometime in the near future the demand for more secure hash functions will arise but these hash functions should also fulfill industry's expectations as long as the throughput ,the area and the power of these new implementations are concerned.In this paper, an implementation of SHA-256 is presented in which the achieved throughput exceeds the limit of 2 Gbps. Furthermore, the proposed implementation results to a lower power dissipation compared to the conventional implementation.

Keywords: Hash-function,SHA-256,High-Throughput,Parallelism

1 Introduction

Nowadays many applications like the Public Key Infrastracture (PKI), IPsec and the 802.16 standard for Local and Metropolitan Area Networksthat incorporate authenticating services. On the other hand applications like SET (Secure Electronic Transactions) have started to concentrate on mobile and portable devices.

All these previously mentioned applications presuppose that an authenticating module that includes a hash function is nested in the implementation of the application.Taking in consideration all different kind of applications that need the services of a hash function we result to the conclusion that all proposed implementations should be optimized in terms of performance, power dissipation and size.

Nowadays the most used hash function is SHA-1.However a collision has been discovered in the full version from the researchers Xiaoyun Wang and Hongbo Yu from Shandong University and Yiqun Lisa Yin from Princeton University.

After this report the US National Institute of Standards and Technology (NIST) has begun recommending that government phase out SHA-1 in favor of SHA-256 and SHA-512.In this paper a high-throughput and low-power implementation of SHA-256 hash function is presented

[1]Authors E-mails: michail@cc.upatras.gr,kakaruda@ee.upatras.gr,milidon@ee.upatras.gr goutis@ee.upatras.gr

2 Proposed Implementation

The Secure Hash Standard [1] describes in detail the SHA-256 hash function. It requires 64 identical operations, to generate the Message Digest.This operation is figured in Fig. 1.

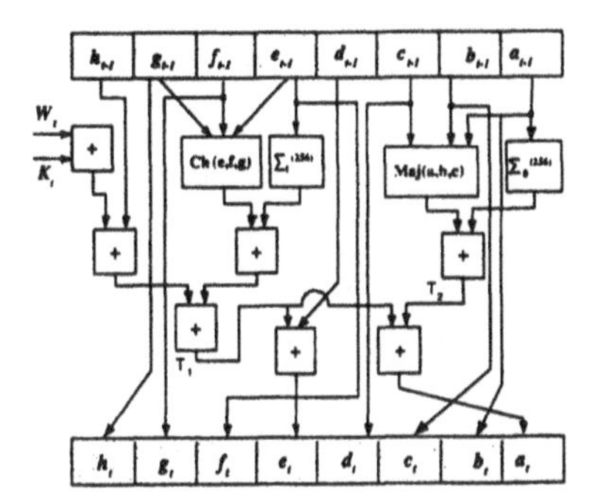

Figure 1: SHA-2 Operation

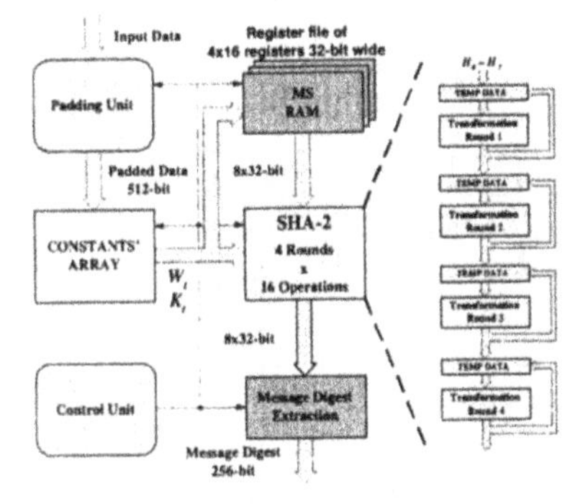

Figure 2: SHA-256 Core Architecture

Each one of the a_t, b_t, c_t, d_t, e_t, f_t, g_t and h_t is 32-bit wide resulting in a 256-bit hash value.K_t and W_t are a constant value for iteration t and the t_{th} w-bit word of the message schedule, respectively. Throughput is kept low due to the large number of the required operations. An approach to increase significantly throughput is the application of pipeline.Applying pipeline the architecture of a SHA-2 core is formed as illustrated in Fig. 2. In the MS RAM, all message schedules W_t of the padded message are stored. The Constants Array is a hardwired array that provides the constant values K_t and the constant initialization values H_0 - H_7. Additionally, it includes the W_t generators.

The proposed design approach is based on a special property of the SHA-2 operation block.Let's consider two consecutive operations of the SHA-2 hash function.The considered inputs a_{t-2}, b_{t-2}, c_{t-2}, d_{t-2}, e_{t-2}, f_{t-2}, g_{t-2} and h_{t-2} go through a specific procedure in two operations and after that the considered outputs a_t, b_t, c_t, d_t, e_t, f_t, g_t and h_t arise.In between the signals a_{t-1}, b_{t-1}, c_{t-1}, d_{t-1}, e_{t-1}, f_{t-1}, g_{t-1} and h_{t-1} exist that are outputs from the first operation and inputs for the second operation.Except of the signal a_{t-1} and e_{t-1} the rest of the signals b_{t-1}, c_{t-1}, d_{t-1}, f_{t-1}, g_{t-1} and h_{t-1} are derived directly from the inputs a_{t-2}, b_{t-2}, c_{t-2}, e_{t-2}, f_{t-2}, g_{t-2} respectively. This means consequently that also c_t, d_t, g_t and h_t can be derived directly from the X_{t-2} inputs.

Furthermore,some calculations during the operation are depended only on the primary operation block's inputs and on intermediate results that are sequentially computed.It seems that some of these calculations can be done in parallel for consecutive operations. In Fig. 3, two consecutive SHA-256 operation blocks are presented according to the conventional approach.In Fig. 4,the proposed implementation is presented in which two consecutive operations have been merged and thus their result is computed in only one clock cycle instead of two.The gray marked areas on Fig. 4 indicate the parts of the proposed SHA-256 operation block that operate in parallel and result to the concurrent computation of the primary operation block's outputs.

Inspecting Fig. 3 and Fig. 4 it is obvious that the critical path in the proposed implementation consists of six addition levels instead of the four addition levels consisting the critical path of the non-concurrent implementation. Although, this fact reduces the maximum operation frequency in the proposed implementation, the throughput is increased significantly since the hash value in the proposed instrumentation is computed in only 32 clock cycles instead of 64 in the non-concurrent implementations.This computations lead to the result that theoretically the throughput of the

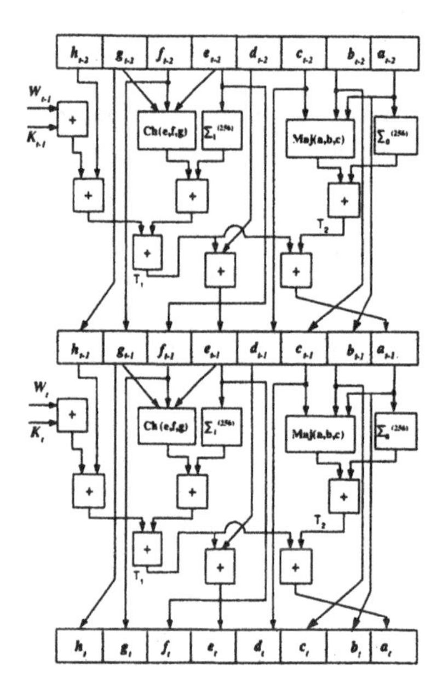

Figure 3: 2 Consecutive Operarions

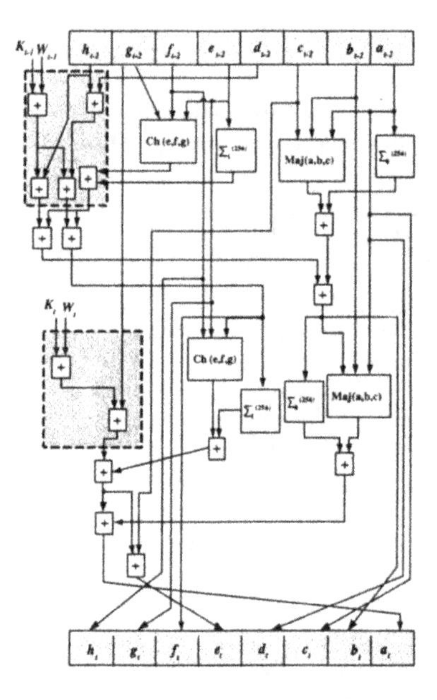

Figure 4: 2 Merged Operations

proposed implementation increases 50%.The experimental result verify this theoretical assumption.

3 Dealing with Power Issues

The decreasing of the operating frequency of the SHA-256 core results to a lower level of dynamic power dissipation for the whole SHA-256 core.This can easily be seen regarding the relevant power equations.Moreover the adopted methodology for the implementation of each SHA-256 operation block combines the execution of two logical SHA-256 operations in only one single clock cycle.This means that the final message digest is computed in only 32 clock cycles and thus calls for only 32 write operations in the temporal register that save all the the intermediate results until the final message digest has been fully derived.

It is possible to estimate the total power savings considering that the initial power dissipation was calculated as P_{init}=$64P_{op}(f_{op}) + 64P_{WR}(f_{op})$, where $P_{op}(f_{op})$ is the dynamic power dissipation of a single operation (depends from the operation frequency f_{op}) and $P_{WR}(f_{op})$ is the power dissipated during write/read operation of the registers (also depends from f_{op}).Both $P_{op}(f_{op})$ and $P_{WR}(f_{op})$'s values are proportional to the operating frequency f_{op}.

According to the latter assumptions, the proposed operation blocks power dissipation is estimated as $P_{prop} = 32(2 * P_{op}(f'_{op})) + 64P_{WR}(f'_{op}) = 64P_{op}(f'_{op}) + 64P_{WR}(f'_{op})$. Considering that $f_{op} > f'_{op}$ and thus $P_{op}(f_{op}) > P_{op}(f'_{op})$ and $P_{WR}(f_{op}) > P_{WR}(f'_{op})$ (according to what was previously mentioned), it can be derived that the operating frequency defines the overall power savings and that the proposed implementation has a lower power dissipation.

Implementations	Operating Frequency(Mhz)	Throughput(Mbps)
[2]	83	326
[3]	74	291
Conv. Impl.	50.1	1632
Prop. Arch.	36.1	2310

Table 1: Operating Frequencies and Throughput

4 Experimental Results-Conclusion

In order to evaluate the proposed SHA-256 design approach, the XILINX FPGA technology was used. The core was integrated to a v150bg352 FPGA device.The operating frequency of the proposed implementation is 36.1 Mhz and the corresponding throughput exceeds 2.3 Gbps.

In Table 1, the proposed implementation and the implementations of [2], [3] and the conventional implementation are compared. From the experimental results, there is a range of 42% - 793% increase of the throughput compared to the previous implementations.

In the case of the introduced area, the implementation of a SHA-256 core,using the proposed operation block, presented a 25% overall area penalty, compared to the conventional operation block implementation. However, the hardware to implement the operation blocks of the SHA-256 rounds is only a small percentage of the whole security scheme. Thus the introduced area is considered to satisfy the requirements of the small-sized SHA-256 implementations.

Furthermore, regarding the overall power dissipation to process a message, the proposed implementation presents significant decrease, approximately by 30% compared to the nearest performing implementation. Power dissipation was calculated using Synopsys Synthesize Flow for the targeted technology and making an aver age estimation for the the activity of the netlist. Then, from the characteristics of the technology, an average wire capacitance was assumed and the power compiler gave rough estimations. Power dissipation is decreased due to the lower operating frequency(without compromising performance) and due to the reduction by 50% of the write processes to the temporal registers. This turns the proposed implementation suitable for portable and mobile devices, meeting this way the constraint for extended autonomy.

Acknowledgment

We thank European Social Fund (ESF), Operational Program for Educational and Vocational Training II (EPEAEK II) and particularly the program PYTHAGORAS, for funding the above work. We would also like to thank Thodoros Giannopoulos for his contribution

References

[1] SHA-2 Standard, National Institute of Standards and Technology (NIST),Secure Hash Standard,FIPS PUB 180-2, www.itl.nist.gov/fipspubs/fip180-2.htm,2002.

[2] N. Sklavos, and O. Koufopavlou, "Implementation of the SHA-2 Hash Family Standard Using FPGAs", Journal of Supercomputing, Kluwer Academic Publishers, Vol. 31, No 3, Issue: X, pp. 227-248, 2005.

[3] N. Sklavos, and O. Koufopavlou, "On the Hardware Implementations of the SHA-2 (256, 384, 512) Hash Functions", proceedings of IEEE International Symposium on Circuits & Systems (ISCAS'03), Vol. V, pp. 153-156, Thailand, May 25-28, 2003.

Brill Academic Publishers
P.O. Box 9000, 2300 PA Leiden,
The Netherlands

Lecture Series on Computer and Computational Sciences
Volume 2, 2005, pp. 174-177

Switchable Data-Dependent Operations: New Designs

Moldovyan N.A. [1], **Moldovyan A.A.**
Specialized Center of Program Systems "SPECTR",
Kantemirovskaya, 10, St.Petersburg 197342, Russia

Introduction

Recently [2, 3, 5] the data-dependent (DD) permutations (DDPs) have been proposed to provide high performance while constrained hardware resources are used. Different types of the DDOs can be implemented using the controlled operations (Definition 1) and defining dependence of the controlling vector on the transformed data.

Definition 1. *Let* $\{\mathbf{F}_1, \mathbf{F}_2, ..., \mathbf{F}_{2^m}\}$ *be some set of the single-type operations defined by formula* $Y = \mathbf{F}_i = \mathbf{F}_i(X_1, X_2, ..., X_q)$, *where* $i = 1, 2, ..., 2^m$ *and* $X_1, X_2, ..., X_q$ *are input n-dimensional binary vectors (operands) and* Y *is the output n-dimensional binary vector. Then the V-dependent operation* $\mathbf{F}^{(V)}$ *defined by formula* $Y = \mathbf{F}^{(V)}(X_1, X_2, ..., X_q) = \mathbf{F}_V(X_1, X_2, ..., X_q)$, *where* V *is the m-dimensional controlling vector, we call the controlled q-place operation. The operations* $\mathbf{F}_1, \mathbf{F}_2, ..., \mathbf{F}_{2^m}$ *are called modifications of the controlled operation* $\mathbf{F}^{(V)}$.

Switchable DDOs (SDDOs) [1] have been proposed as a new primitive. Implementation results shows the SDDO-based ciphers provide high performance while implemented in cheap hardware [4]. The SDDOs are performed with switchable controlled operations (SCO) defined below.

Definition 2. *Let* $\{\mathbf{F}_1, \mathbf{F}_2, ..., \mathbf{F}_{2^m}\}$ *be the set of the modifications of the controlled operation* $\mathbf{F}^{(V)}$. *The operation* $(\mathbf{F}^{-1})^{(V)}$ *containing modifications* $\mathbf{F}_1^{-1}, \mathbf{F}_2^{-1}, ..., \mathbf{F}_{2^m}^{-1}$ *is called inverse of* $\mathbf{F}^{(V)}$, *if* $\mathbf{F}_V^{-1}$ *and* $\mathbf{F}_V$ *for all* V *are mutual inverses.*

Definition 3. *Let* $\mathbf{F}'^{(e)}$, *where* $e \in \{0,1\}$, *be some e-dependent operation containing two modifications* $\mathbf{F}'^{(0)} = \mathbf{F}'_1$ *and* $\mathbf{F}'^{(1)} = \mathbf{F}'_2$, *where* $\mathbf{F}'_2 = \mathbf{F}'^{-1}_1$. *Then the operation* $\mathbf{F}'^{(e)}$ *is called switchable.*

Definition 4. *Let two modifications of the switchable operation* $\mathbf{F}'^{(e)}$ *be mutual inverses* $\mathbf{F}'^{(0)} = \mathbf{F}^{(V)}$ *and* $\mathbf{F}'^{(1)} = (\mathbf{F}^{-1})^{(V)}$. *Then* $\mathbf{F}'^{(e)}$ *is called switchable controlled operation* $\mathbf{F}^{(V,e)}$.

In present paper we consider new designs of SDDOs providing cheaper implementation. Troughout the paper we use the following notations:

◇ $\{0,1\}^n$ denotes the set of all n-bit binary vectors $X = (x_1, ..., x_n)$, where $x_i \in \{0,1\}$, $i = 1, ..., n$;

◇ $(A, B, ..., Z)$ denotes concatenation of the binary vectors A, B,..., Z;

◇ "$+_n$" ("$-_n$") denotes addition (subtraction) modulo 2^n; $[s/2]$ denotes integer part of $s/2$.

◇ $Y = X^{>>>k}$ denote rotation of the word X by k bits, where $\forall i \in \{1, ..., n-k\}$ we have $y_i = x_{i+k}$ and $\forall i \in \{n-k+1, ..., n\}$ we have $y_i = x_{i+k-n}$.

Networks based on minimum size controlled elements

The known controlled operations (CO) are implemented as uniform controlled substitution-permutation networks (CSPNs) constructed using some controlled element (CE) of minimum size as standard building block. Figure 1a shows general topology of CSPN. Let SCPN with n-bit input and n-bit

[1]Corresponding author. E-mail: nmold@cobra.ru

output be controlled with m-bit vector V. Then we shall denote such SCPN as controlled operation (CO) $\mathbf{F}_{n;m}$. Selecting a set of the fixed permutations connecting active layers we define some particular topology of CO. Each active layer represents $n/2$ parallel CEs. The $\mathbf{F}_{2;1}$ minimum size CE (Fig. 1b) transforms two-bit input vector (x_1, x_2) into two-bit output (y_1, y_2) depending on some controlling bit v. The CE can be described with a pair of BFs in three variables (Fig. 1b) : $y_1 = f_1(x_1, x_2, v)$; $y_2 = f_2(x_1, x_2, v)$.

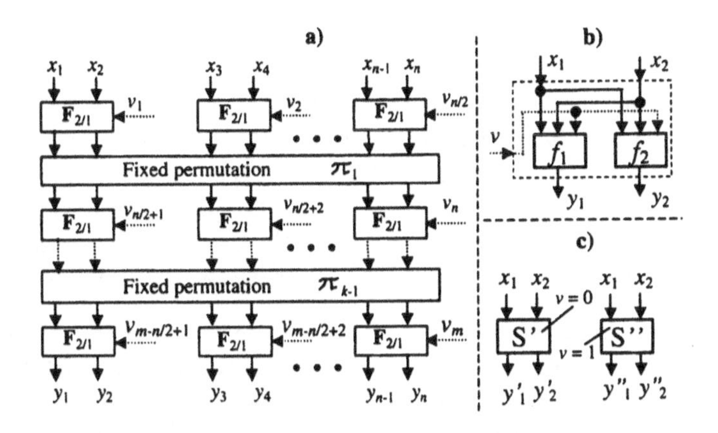

Figure 1: General structure of the $\mathbf{F}_{n;m}$ boxes (a) and representation of the $\mathbf{F}_{2;1}$ element as a pair of BFs in three variables (a) or as switchable 2×2 substitution (b)

A CE can be also represented as a pair of the 2×2 substitutions selected depending on bit v (Fig. 1c). Such subsitutions are denoted as $\mathbf{F}_{2;1}^{(0)}$ and $\mathbf{F}_{2;1}^{(1)}$ and CE implements the transformation $(y_1, y_2) = \mathbf{F}_{2;1}^{(v)}(x_1, x_2)$. The $\mathbf{F}_{n;m}$ box can be represented as a superposition of the operations performed on binary vectors:

$$\mathbf{F}_{n;m} = \mathbf{L}^{(V_1)} \circ \pi_1 \circ \mathbf{L}^{(V_2)} \circ \pi_2 \circ \ldots \circ \pi_{s-1} \circ \mathbf{L}^{(V_s)},$$

where π_j, $j = 1, 2, \ldots, s-1$, are fixed permutations, $V = (V_1, V_2, \ldots, V_s)$, V_j is the component of V, which controls the jth active layer, and $s = 2m/n$ is the number of active layers $\mathbf{L}^{(V_j)}$.

Suppose a CO box is constructed using CEs that are involutions. Then one can easy construct the layered box $\mathbf{F}_{n;m}^{-1}$ which is inverse of $\mathbf{F}_{n;m}$-box:

$$\mathbf{F}_{n;m}^{-1} = \mathbf{L}^{(V_s)} \circ \pi_{s-1}^{-1} \circ \mathbf{L}^{(V_{s-1})} \circ \pi_{s-2}^{-1} \circ \ldots \circ \pi_1^{-1} \circ \mathbf{L}^{(V_1)}.$$

In accordance with the structure of the CO boxes $\mathbf{F}_{n;m}$ and $\mathbf{F}_{n;m}^{-1}$ we shall assume that in the direct CO boxes the $\mathbf{F}_{2;1}$ elements are consecutively numbered from left to right and *from top to bottom* and in the inverse CO boxes the CEs are numbered from left to right and *from bottom to top*. Thus, for all $i \in \{1, 2, \ldots, m\}$ the ith bit of the controlling vector V controls the ith box $\mathbf{F}_{2;1}$ in both the $\mathbf{F}_{n;m}$ and $\mathbf{F}_{n;m}^{-1}$ boxes. For $j = 1, 2, \ldots, s$ the V_j component of V controls the j-th active layer in the box $\mathbf{F}_{n;m}$ and the $(s-j+1)$-th layer in $\mathbf{F}_{n;m}^{-1}$. Below we will use the following definition ($\mathbf{L}_j$ denotes the jth active layer):

Definition 5. *The CO box $\mathbf{F}_{n;m}$ is called symmetric if $\forall j = 1, \ldots, s-1$ the following relations are hold: $\mathbf{L}_j = \mathbf{L}_{s-j+1}^{-1}$ (or $\mathbf{L}_j = \mathbf{L}_{s-j+1}$, if $\mathbf{L}_j$ is involution) and $\pi_j = \pi_{s-j}^{-1}$.*

Switchable data-dependent operations

Earlier [1] the symmetric SCOs $\mathbf{F}^{(V,e)}_{32;96}$ and $\mathbf{F}^{(V,e)}_{64;192}$ has been considered as new primitive suitable to the design of the 64- and 128-bit ciphers, correspondingly. These SCOs are based on i) the use of the six-layer CSPNs $\mathbf{F}^{(V)}_{32;96}$ and $\mathbf{F}^{(V)}_{64;192}$ having mirror-symmetry topology and ii) swapping the V_j and V_{7-j} components of the controlling vector V for $j = 1, ..., 6$. Due to symmetric structure of CO boxes the modifications $\mathbf{F}^{(V)}$, where $V = (V_1, V_2, ...V_6)$, and $\mathbf{F}^{(V')}$, where $V' = (V_6, V_5..., V_1)$ are mutually inverse. In the $\mathbf{F}^{(V,e)}_{32;96}$ box swapping the V_j and V_{7-j} components is performed with very simple transposition box $\mathbf{P}^{(e)}_{96;1}$ that is implemented as some single layer CP box consisting of three parallel single-layer boxes $\mathbf{P}^{(e)}_{2\times 16;1}$. Input of each $\mathbf{P}^{(e)}_{2\times 16;1}$-box is divided into 16-bit left and 16-bit right inputs. The box $\mathbf{P}^{(e)}_{2\times 16;1}$ represents 16 parallel $\mathbf{P}^{(e)}_{2;1}$-boxes controlled with the same bit e. The right (left) input (output) of 16 parallel boxes $\mathbf{P}^{(e)}_{2;1}$ compose the right (left) 16-bit input (output) of the box $\mathbf{P}^{(e)}_{2\times 16;1}$. Thus, each of three boxes $\mathbf{P}^{(e)}_{2\times 16;1}$ performs e-dependent swapping of the respective pair of the 16-bit components of the controlling vector V. If the input vector of the box $\mathbf{P}^{(e)}_{96;1}$ is $(V_1, V_2, ...V_6)$, then at the output of $\mathbf{P}^{(e)}_{96;1}$ we have $V' = (V_1, V_2, ..., V_6)$ (if $e = 0$) or $V' = (V_6, V_5, ..., V_1)$ (if $e = 1$).

The $\mathbf{F}^{(V,e)}_{64;192}$ box can be constructed with the use of transposition box $\mathbf{P}^{(e)}_{192;1}$ that represents a cascade of three single-layer boxes $\mathbf{P}^{(e)}_{2\times 32;1}$. Each $\mathbf{P}^{(e)}_{2\times 32;1}$-box is a set of 32 parallel $\mathbf{P}^{(e)}_{2;1}$-boxes all of which are controlled with the bit e. The SCO design considered above imposes no restrictions on the distribution of the controlling bits, however in the DDO-based ciphers the m-bit controlling vector V depends usually on some n-bit controlling data subblock L. The V vector is formed as output of the extension operation $\mathbf{E}$ performed on L. Since we have $m = 3n$ one can propose SCO construction with reduced implementation cost designing respective mechanism of swapping some components of L. We propose the following design.

Let L_1 and L_2 be two halfs of $L = (l_1, l_2, ..., l_n)$, i. e. $L = (L_1, L_2)$. The L_1 and L_2 components are extended using two symmetric extension boxes $\mathbf{E}_1$ and $\mathbf{E}_2$. The outputs of $\mathbf{E}_1$ and $\mathbf{E}_2$ are $V' = (V_1, V_2, V_3)$ and $V'' = (V_4, V_5, V_6)$, respectively. Thus, the boxes $\mathbf{E}_1$ and $\mathbf{E}_2$ represent a single extention box $\mathbf{E}$ with symmetric structure, which produces the controlling vector $V = (V', V'')$ corresponding to symmetric distribution of the bits of the $L_1 = (l_1, l_2, ..., l_{n/2})$ and $L_2 = (l_{n/2+1}, l_{n/2+2}, ..., l_n)$ components. In this case swapping the L_1 and L_2 components defines swapping components V_j and V_{7-j} for $j = 1, ..., 6$. For the $\mathbf{F}^{(L,e)}_{32;96}$ operation we propose the $\mathbf{E}$ box having the following structure:

$$V_1 = L_1;\ \ V_2 = L_1^{>>>6};\ \ V_2 = L_1^{>>>12};\ \ V_4 = L_2^{>>>12};\ \ V_5 = L_2^{>>>6};\ \ V_6 = L_2.$$

For the $\mathbf{F}^{(L,e)}_{64;192}$ operation we propose the $\mathbf{E}$ box described as follows:

$$V_1 = L_1;\ \ V_2 = L_1^{>>>14};\ \ V_2 = L_1^{>>>28};\ \ V_4 = L_2^{>>>28};\ \ V_5 = L_2^{>>>14};\ \ V_6 = L_2.$$

New desing of the SCO boxes $\mathbf{F}^{(L,e)}_{32;96}$ and $\mathbf{F}^{(L,e)}_{64;192}$ is presented in Fig. 2a,b. Thus, to embed the switchability mechanism in the $\mathbf{F}_{32;96}$ and $\mathbf{F}_{64;192}$ boxes we need only 96 and 192 extra nand gates, correspondingly, versus 288 and 576 extra nand gates for the design proposed in [1].

Other hardware efficient design of the SCO boxes is based on the use of CEs that implements mutually inverse modifications $\mathbf{F}^{(0)}_{2;1}$ and $\mathbf{F}^{(1)}_{2;1}$ for which we have $\mathbf{F}^{(0)}_{2;1} = \left(\mathbf{F}^{(1)}_{2;1}\right)^{-1}$. In this design the symmetric topology of the CO boxes combined with symmetric distribution of the controlling bits (see Definition 6) is used. Figure 2c,d, where $E(e) = \{e\}^{2k}$ is concatenation of $2k$ bits all of which are equal to e, illustrates the construction of the SCO boxes $\mathbf{F}^{(L,e)}_{32;96}$ (c) and $\mathbf{F}^{(L,e)}_{64;192}$ (d) (topology of

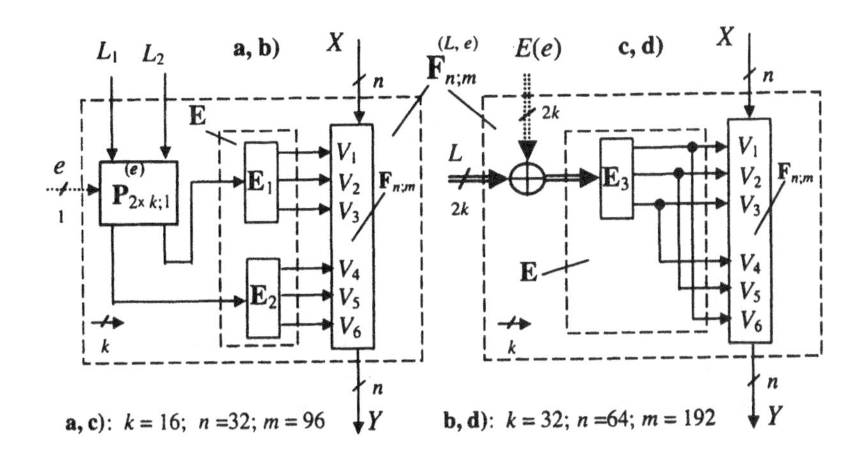

Figure 2: The $\mathbf{F}^{(L,e)}_{32;96}$ (a,c) and $\mathbf{F}^{(L,e)}_{64;192}$ (b,d) boxes with reduced implementation cost: a,b) - switching by swapping the left and right halves of the controlling data subblock; c,d) - switching by inverting the controlling bits

the $\mathbf{F}_{32;96}$ and $\mathbf{F}_{64;192}$ boxes is described in [1]). It is easy to show that this construction provides the relation $\mathbf{F}^{(L,e)}_{n;m} = \left(\mathbf{F}^{(L,e\oplus 1)}_{n;m}\right)^{-1}$.

Definition 6. *Distribution of the controlling bits of the* $V = V_1, ..., V_{[s/2]}$ *vector is called symmetric if* $\forall j = 1, ..., [s/2]$, *the following relation is hold:* $V_j = V_{s-j+1}$.

References

[1] N. A. Moldovyan, On Cipher Design Based on Switchable Controlled operations. *Springer-Verlag LNCS,* vol. 2776 (2003), pp. 316-327.

[2] A.A.Moldovyan and N.A.Moldovyan, A cipher based on data-dependent permutations, *Journal of Cryptology* vol. 15, no. 1 (2002), pp. 61-72.

[3] N. Sklavos, N.A. Moldovyan, O. Koufopavlou, High Speed Networking Security: Design and Implementation of Two New DDP-Based Ciphers, *Mobile Networks and Applications,* 2005, vol. 10, no. 1, pp. 237-249

[4] N. Sklavos and O. Koufopavlou, Architectures and FPGA Implementations of the SCO (-1,-2,-3) Ciphers Family. *Proceedings of the of 12th International Conference on Very Large Scale Integration, (IFIP VLSI SOC '03),* Darmstadt, Germany, December 1-3, 2003.

[5] N. Sklavos, and O. Koufopavlou, "Data Dependent Rotations, a Trustworthy Approach for Future Encryption Systems/Ciphers: Low Cost and High Performance", Computers and Security, Elsevier Science Journal, Vol. 22, Number 7, pp. 585-588, 2003.

Brill Academic Publishers
P.O. Box 9000, 2300 PA Leiden,
The Netherlands

Lecture Series on Computer and Computational Sciences
Volume 2, 2005, pp. 178-183

A Low Power, High Frequency and High Precision Multiplier Architecture in GF(P)

Abdoul Rjoub[†] and Lo'ai A. Tawalbeh
Department of Computer Engineering
Jordan University of Science and Technology

Abstract: A high bit modular multiplier based on novel design of Radix-4 Montgomery architecture in the layout design is presented in this paper. It is based on using two types of digit recoding which results in a competitive and increased in speed operation of the design, smallest transistor accounts and the power/speed ratio. We showed in previous publications that radix-4 multiplier operates at high speed and has relatively small area which makes it one of the best solutions in many arithmetic-dependent applications, such as public-key cryptography algorithms and signal processing applications. In this paper, we are investigating the power dissipation of the proposed architecture and we are presenting the hardware components using transmission gates which reduce the power dissipation at minimum value. The new design is implemented in 0.12-μm double-poly double-metal CMOS process, with 1.2 V power supply.

Keywords: Public-key cryptographic, High speed multiplier, Low power design, Layout.

1. Introduction

The high-precision multipliers over prime finite fields (GF(p)) form the basic components in arithmetic-based applications such as public-key cryptography algorithms [1][2]. For instance, the Elliptic Curve crypto-systems [3][4] and the Diffie-Helman key exchange [5] use wide-width operands to provide a reasonable level of security, and as a result, a huge amount of multiplications have to be executed. So, speeding up these computations, speeds up the whole design. In other words, the multiplier design has a main role in determining the performance of the overall system, and so, in order to have an efficient system, the designer should take in consideration designing a fast and efficient multiplier.

The radix-4 arithmetic circuits provide best solutions in many cases [6]. The multiplier design obtained in this paper is based on Radix-4 Montgomery multiplier [7]. The multiplier is used for prime finite fields (GF(p)). The architecture uses two types of digit recoding. The first is Booth recoding for the multiplier digits to get rid of the multiple "3" form the critical path. The second recoding is for the multiplies of the modulus (Montgomery algorithm requires addition of the modulus at certain points [7][8]). Using these two types of digit recoding, results in a high frequency multiplier design that is suitable for cryptographic hardware implementations. Also, the carry-save number system is used in the design to perform the additions independently from the operands precision, especially, if we are designing a wide-bit multiplier which is the case in most of cryptographic algorithms.

For a design to be considered efficient, the power dissipation as well as circuit speed and area should be considered at the architectural level. Recently, more focus on reducing the power dissipation of these important designs has been placed, with maintaining high peed operation. Reducing the power dissipation in today's synthesis tools is done using built-in libraries. These libraries use traditional

[†] E-mail: abdoul,tawalbeh@just.edu.jo

CMOS components in building a specific design which might not be up-to-date with the modern techniques in power reduction and performance optimization. One approach to minimize power dissipation is to layout the design from the scratch by hand. In other words, designing each component using the appropriate number of low-power building blocks (transmission gates in our case) and then integrating the whole design by hand. This is time-consuming process for large designs, but on the other hand, optimal power designs are obtained.

In this paper, the Radix-4 multiplier proposed in [7] is designed from the scratch in the layout level to satisfy minimum power dissipation. In order to achieve this target, we implemented each component in the design (Carry Save Adders, Booth Encoders, and the basic logic gates such as AND and XOR gates) using multiplexers to reduce the dynamic
power dissipation which is the main factor in power dissipation in CMOS circuits. The performance of the new design is assisted from many aspects such as the delay and the transistor count.

The following Section presents a low-power design of each component in the multiplier architecture. Each component built using only multiplexers (MUXs) to get minimum power dissipation. The delay, transistor count, and power characteristics of the proposed multiplier architecture is discussed in Section 3. Section 4 concludes this paper.

2. Radix-4 Low-power Multiplier Components

We showed in previous publications [7] multiplier operates at high speed and has relatively small area which makes it the best solution in designing many specialized systems such as the public-key cryptography applications. Figure 1 shows the Radix-4 multiplication cell. The main components are the Carry-Save Adders (CSA) and the Booth encoder and the Modulus encoder. Registers also used to store the intermediate values.

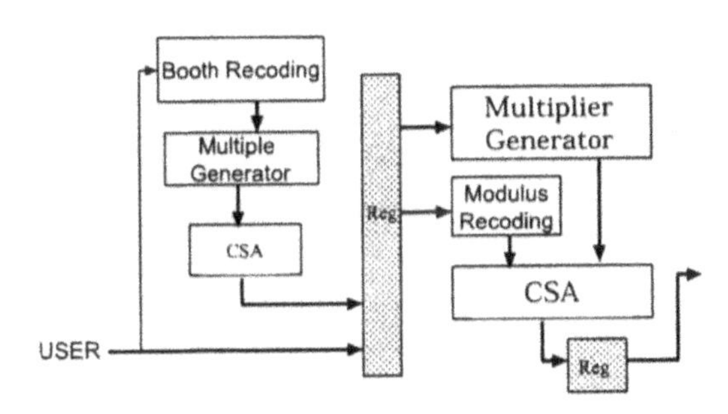

Figure 1: Radix-4 Montgomery Multiplication Cell

The Multiple Generators are composed form basic logic gates.

In this section, we design each component of the multiplier in a way so that it dissipates minimal power. We are using multiplexers (MUXs) as the only building blocks in each component. The multiplexers consist of transmission gates and one small-area inverter. In our design, a 2:1 multiplexer consists of two transmission gates, and it is used to replace the basic traditional CMOS gates such as AND, OR, and XOR.

The figures of the components are generated using the *MicroWind 3* tool. Figure 2 shows one bit carry-save adder designed using multiplexers instead of XOR and other COMS gates.

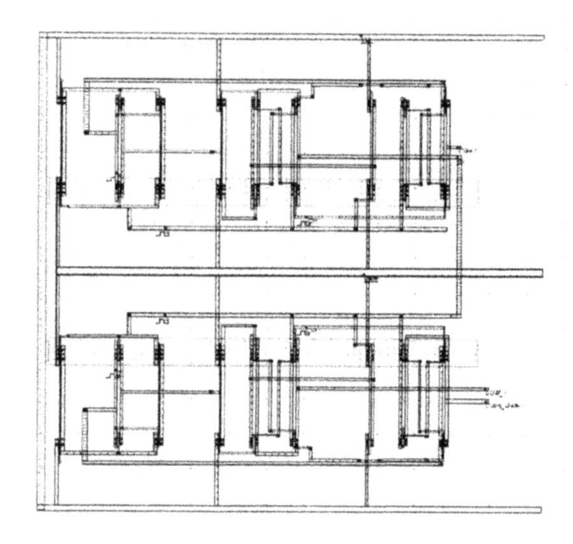

Figure 2: One-bit Carry-Save Adder layout.

Figure 3 shows the Booth recoding layout implemented using only multiplexers.

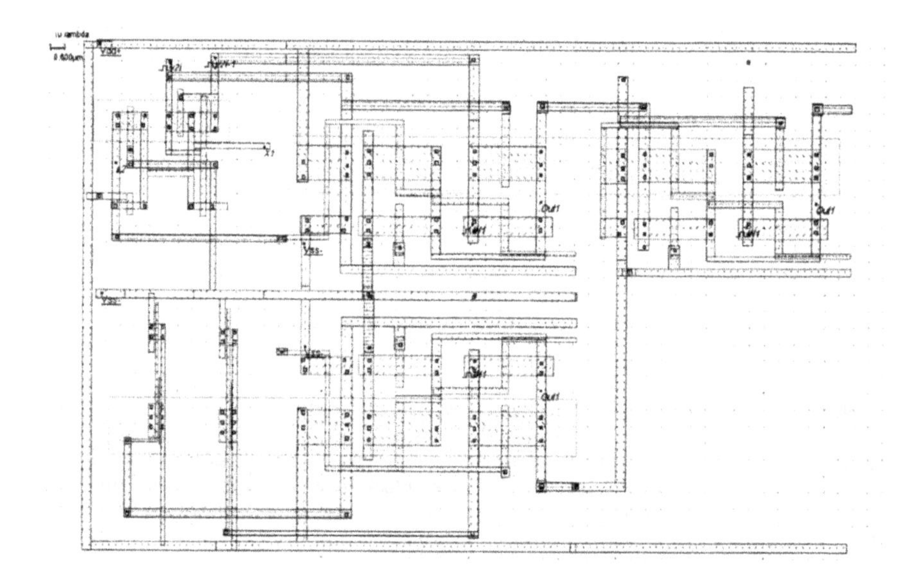

Figure 3: The Booth Recoding Layout.

The modulus recoding block is just a lookup table and it is not in the critical path. The multiple generators consists of a (2:1 MUX) and an XOR gate which are laid out using transmission gates.
The registers layout is generated by the same way, and it is shown in Figure 4 below:

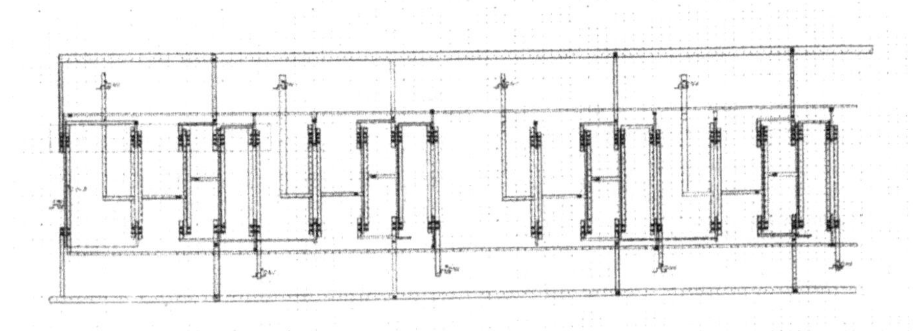

Figure 4: The Register Layout.

We will investigate the power and delay properties and the number of transistors of each component in the next Section.

3. Performance Analysis

In this section we study the delay, transistor count and the power dissipation characteristics of the 128x128 bit radix-4 multiplier. The results in this Section are obtained by the *MicroWind 3* Layout tools using CMOS 0.12 μ meter, 1.2 V, 6 metals. The delay presented in subsection 3.1 is obtained for 1-bit multiplier and calculated for 128 bit. In subsection 3.2 we present the number of transistors in each component and calculate it for 1-bit multiplier and then for 128 bit multiplier. Finally, in 3.3 we show the power dissipation characteristics of the 128-bit multiplier.

3.1 Delay

Assuming that the delay of a component is T_component. For example, T_CSA is the delay of the Carry Save Adder. Table 1 shows the delay in picoseconds (ps) for the main components in the multiplier (each of one bit width):

Table 1: Time Delay of the Main Components in the Multiplier

Delay		
T_CSA	**T_Booth**	**T_Register**
30 ps	**20 ps**	**11 ps**

The critical path delay of the multiplier (let us call it T_mult) is determined by:

$$T_{mult} = T_{Booth} + T_{Multiple\_Generator} + 2\times T_{\_Register} + 2T_{\_CSA}$$

Consider

$$T_{Multiple\_Generator} = T_{XOR} + T_{2:1Mux}) = 2T_{2:1Mux}) = 2\times 9 = 18ps$$

then:

$T_{mult} = 107\,ps$ is the delay of one-bit multiplier.

The 128×128 bit multiplier will have a delay of 128×128 = 15.36 ns

By the same way, the delay can be computed for any nxn multiplier.
We have to say that the delay obtained by this way is an approximation since in the real lay out, other factors should be considered.

3.2 Transistor Count

Table 2 shows the number of transistors in the main components of the multiplier:

Table 2: Transistor Count of The Main Components in The Multiplier

Transistor Count		
A_CSA	A_Booth	A_Register
36	27	10

Where A_component denoted the area of a specific component. Considering a 2:1 MUX has an area of around 6 transistors, and each multiple generator consists of two MUXs, then the total area of a one bit radix-4 multiplier is given by:

$$A_{mult} = T_{Booth} + 2 \times T_{Multiple\_Generator} + 2 * T_{Register} + 2 * T_{Register} + 2T\_CSA = 143 \text{ transistors}$$

We remind the reader that these numbers are approximations since the multiplier design has a look up table component which was not included in the calculations. The number of transistors in a 128x128-bit radix-4 multiplier will be = 143x128 =18,304 transistors.

3.3 Power Dissipation

Table 3 shows the power dissipation in mWatt for the main components in the radix-4 multiplier:

Table 3: Power Dissipation of the Main Components in The Multiplier

Power Dissipation (mW)		
P_CSA	P_Booth	P_Register
5	5	2

Where the power dissipated by a component is denoted by P_component.

Considering a 2:1 mux dissipates 0.1 mWatt of power, then the total power dissipation of the radix-4 multiplier will be approximately:

P_mult = 25 mW.

The total power dissipated by 128x128-bit radix-4 multiplier will be = 25x128 =3200 mW.

From the above results, we can see that the radix-4 multiplier constructed in this paper has a low power dissipation and a small area and still operates at high speeds and it can be used for large-precision operands.

4. Conclusion

In this paper we obtained a low-power Radix-4 multiplier by designing it's hardware components using transmission gates (2:1 Muxs). The complete layout of the complete multiplier design was generated by hand for minimal power dissipation. The design has small area and maintained high speed operation, and it is used for high precision operands. This makes it suitable for hardware implementations of public-key cryptography algorithms.

References

[1] I. Kourdoulis, N. Sklavos, P. Kitsos and O. Koufopavlou," VLSI Design and Implementation of RSA Encryption Algorithm," Proceedings of International Arab Conference on Information Technology (ACIT'01), pp. 299-302, Jordan, November 13-15, 2001.

[2] Nicolas Sklavos and Odysseas Koufopavlou, " Mobile Communications World:

Security Implementations Aspects - A State of the Art," *CSJM Journal.*

[3] N. Koblitz, "*Elliptic curve cryptosystems*," Mathematics of computation, vol. 48, no. 177, pp. 203-209, January 1987.

[4] A. J. Menezes, "*Elliptic curve public key cryptosystems*," Kluwer Academic Publishers, Bosten, MA. USA, 1993

[5] M. E. Hellman and W. Diffie, "*New directions on cryptography*," IEEE transactions on Information Theory, vol. 22, pp. 644-654, November 1976.

[6] B. S. Cherkauer and E. B. Friedman, "*A hybrid radix-4/radix-8 low power, high speed multiplier architecture for wide bit widths*," In the IEEE International Symposium on Circuits and Systems. 1996.

[7] A. F. Tenca and L. A. Tawalbeh, "*An efficient and scalable radix-4 modular multiplier design using recoding techniques*," in The Thirty-seventh Annual Asilomar Conference on Signals, Systems, and Computers. November 9-12, 2003, vol. 2, pp. 1445{1450, IEEE Press, Pacific Grove, California.

[8] A. F. Tenca and C. K. Koc, "*A word-based algorithm and scalable architecture for Montgomery multiplication*," in Cryptographic Hardware and Embedded Systems-CHES 1999, C. K. Koc and C. Paar, Eds. 1999, Lecture Notes in Computer Science, No. 1717, pp. 94{108, Springer, Berlin, Germany.

Brill Academic Publishers
P.O. Box 9000, 2300 PA Leiden,
The Netherlands

Lecture Series on Computer and Computational Sciences
Volume 2, 2005, pp. 184-190

A New 64-Bit Cipher for Efficient Implementation Using Constrained FPGA Resources

[A] N. A. Moldovyan[1], [A] A. A. Moldovyan, and [B] N. Sklavos

[A] Specialized Center of Program Systems, SPECTR, Kantemirovskaya Str. 10,
St. Petersburg 197342, Russia, email: nmold@cobra.ru

[B] Electrical & Computer Engineering Department,
University of Patras, Greece, email: nsklavos@ee.upatras.gr

Abstract: This work focuses the problem of increasing the integral VLSI implementation efficacy of block ciphers, which are suitable to applications where constrained resources are available to embedded security mechanisms in both ad-hoc and sensor networks. In this paper a new 64-bit cipher Eagle64 is developed, based on the approach of Data Dependent (DD) Operations (DDOs). The proposed Eagle64 provides very fast encryption, while constrained FPGA resources are utilized. The implementation synthesis results are compared with other well known ciphers implementations. It is shown the proposed cipher hardware integration is more efficient than conventional 64- and 128-bit ciphers, using the ratio of both "Performance/Cost" and "Performance/(Cost·Frequency)".

Keywords: Data Dependent Operations, Eagle64, Hardware Implementation, Block Ciphers

1. Introduction

Security and Privacy has attracted the research community interest the last years, especially in the field of ad-hoc and sensor networks, where only constrained hardware resources are often available. Especially privacy is often provided using data encryption and designing new block ciphers well suited to the efficient hardware implementation has significant practical importance. Encryption algorithms have to perform efficiently in a variety of current and future applications, doing different encryption tasks. Optimizations of the existing security standards as well as novel designs are proved issues of major importance in order the high needs for security to be satisfied. A new approach to the block cipher design is based on the use of the data-dependent operations (DDOs). It has been earlier proposed in [1,2] as a trustworthy and flexible solution to such practical problems.

Usually the DDO boxes are implemented as controlled substitution-permutation networks (CSPNs). Data-Dependent (DD) Permutations (DDPs) represent a particular case of DDOs. The DDP boxes are implemented with permutation networks (PNs) [2]. The efficiency of the DDP as cryptographic primitive is caused by the variability of this operation while ciphering different data blocks. Security estimations [3-5] and implementation synthesis results [6-8] prove the efficiency of the DDP-based design, the DDP boxes are linear cryptographic primitive though. The next step to advance the DDO-based design has been presented in [9], where the CSPNs with minimum size controlled elements (CEs) have been proposed as DDP-like operations, which are free of any linear characteristics. In both the DDP and the DDP-like boxes, the CEs are used as standard building blocks. The last implement one of two possible 2×2 substitutions depending on the one controlling bit. Such CEs are denoted as $\mathbf{F}_{2/1}$.

[1] Corresponding author

Recently [10] it has been shown that there exist 128 different types of non-linear CEs and all of them can be used to design non-linear DDO boxes. Besides, such operational boxes contribute to the avalanche significantly better than DDP boxes. Since the CEs of all types are implemented in FPGA using two configurable logic blocks (CLBs) the non-linear CEs are preferable to design DDOs that have advanced properties. In [11] it has been noted that while implementing a CE only 50% of the resources of two CLBs are utilized. The paper proposes another step advancing the DDO-based design that is to use the minimum size CEs denoted as $\mathbf{F}_{2/2}$ and controlled with two bits instead of the CEs controlled with one bit. In this way, 100% of the resources of two CLBs are used. In the new type CEs each output is a Boolean function (BF) in three variables having higher non-linearity (in the sense of the dinstance from the set of BFs in the same number of variables) value NL than BFs in three variables, which describe the $\mathbf{F}_{2/1}$ non-linear elements.

In general a DDO box having n-bit input and constructed using the $\mathbf{F}_{2/2}$ element as standard building block is shown in Fig. 1, where the controlled vector (V) of the box is represented as concatenation of $2s$ components V_i and Z_i ($i = 1, \quad , s$) of the ($n/2$)-bit length: $V = (V_1, Z_1, \ldots, V_s, Z_s)$.

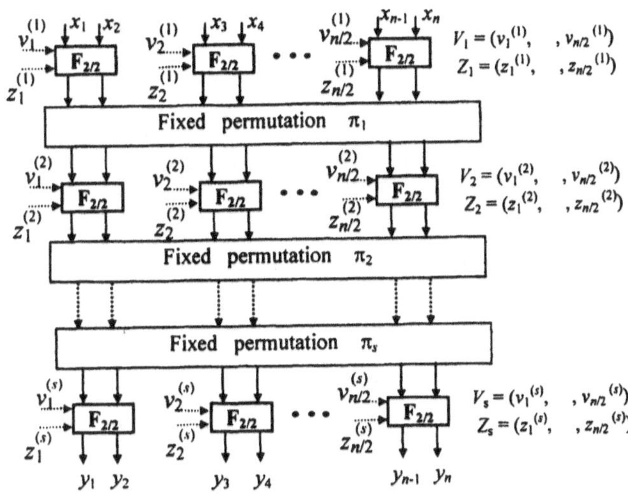

Figure 1: Controlled substitution-permutation network built up using the minimum size CEs $\mathbf{F}_{2/2}$

Selecting different fixed permutations $\pi_1, \quad , \pi_{s-1}$ different types of the DDO boxes can be constructed, using the single type CE $\mathbf{F}_{2/2}$. For example, the fixed permutations applied in the DDP operations design [1,7], could be used. The $\mathbf{F}_{2/2}$ elements (Fig. 2a) can be described in two different ways: 1) as a pair of two BF in four variables (Fig. 2b) and 2) as a set of four 2×2 substitutions (S_1, S_2, S_3, S_4) (Fig. 2c). The first representation can be get from the second one using the following formula:

$$y_1 = vz(y_1^{(1)} \oplus y_1^{(2)} \oplus y_1^{(3)} \oplus y_1^{(4)}) \oplus v(y_1^{(1)} \oplus y_1^{(3)}) \oplus z(y_1^{(1)} \oplus y_1^{(2)}) \oplus y_1^{(1)}$$
$$y_2 = vz(y_2^{(1)} \oplus y_2^{(2)} \oplus y_2^{(3)} \oplus y_2^{(4)}) \oplus v(y_2^{(1)} \oplus y_2^{(3)}) \oplus z(y_2^{(1)} \oplus y_2^{(2)}) \oplus y_2^{(1)}$$

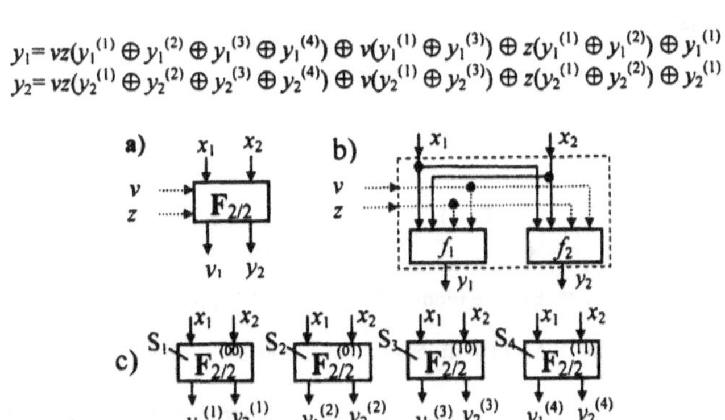

Figure 2: Element $\mathbf{F}_{2/2}$ (a) represented as a pair of BFs (b) or as four 2×2 substitutions (c)

It has been shown that there exist 2208 different variants of the $\mathbf{F}_{2/2}$ non-linear elements having the following properties:

1. *Each of two outputs of CEs is a balanced BF (in four variables) having non-linearity NL = 4.*
2. *Each modification of CEs is a bijective transformation $(x_1, x_2) \rightarrow (y_1, y_2)$ that is an involution.*
3. *The sum of two outputs, i.e. $f_3 = y_1 \oplus y_2$, is a BF (in four variables) having non-linearity NL = 4.*

In this work we select the $\mathbf{F}_{2/2}$ elements that provide construction of the DDO boxes contributing well to the avalanche. Furthermore, based on them the design a new cipher Eagle64 is proposed, which is proven significantly more efficient for FPGA implementation than other conventional 64-bit ciphers.

2. Design of the DDO boxes

Except the NL value and algebraic degree of BF, differential characteristics (DCs) of the CE are of great importance, to characterize CEs as cryptographic primitives. In order toselect the $\mathbf{F}_{2/2}$ CEs suitable to designing efficient cryptographic DDOs, differential characteristics DCs (see Fig. 3) of all 2208 elements having maximum non-linearity $(\mathrm{NL}(y_1) = \mathrm{NL}(y_2) = \mathrm{NL}(y_1 \oplus y_2) = 4)$, have been investigated.

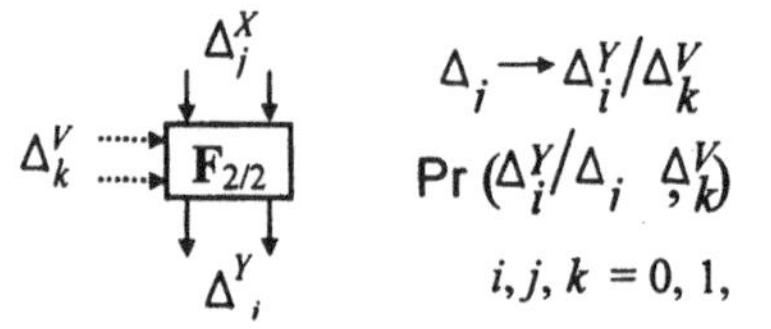

Δ_j^X Δ_k^V $\mathbf{F}_{2/2}$ Δ_i^Y $\Delta_j \rightarrow \Delta_i^Y/\Delta_k^V$ $\Pr(\Delta_i^Y/\Delta_j, \Delta_k^V)$ $i, j, k = 0, 1,$

Figure 2: Differential characteristics of the $\mathbf{F}_{2/2}$ elements

To characterize DCs the integral parameter, called average entropy, has been used. This is defined as follows:

$$\overline{\mathrm{H}} = \frac{\left(\sum_{j=0}^{2}\sum_{k=1}^{2}\mathrm{H}_{jk} + \sum_{j=1}^{2}\mathrm{H}_{j0}\right)}{8}, \quad \text{where} \quad \mathrm{H}_{jk} = -\sum_{i=0}^{2} p\left(\Delta_i^Y/\Delta_j^X, \Delta_k^V\right)\log_3 p\left(\Delta_i^Y/\Delta_j^X, \Delta_k^V\right)$$

According to our research 128 different variants of the $\mathbf{F}_{2/2}$ elements for which we have $\overline{\mathrm{H}} = 0.840$ (this is maximal value). Additionally, there exist four other subclasses for which we have $\overline{\mathrm{H}} > 0.8$: i) 704 variants with $\overline{\mathrm{H}} = 0.834$, ii) 128 variants with $\overline{\mathrm{H}} = 0.815$, iii) 192 variants with $\overline{\mathrm{H}} = 0.813$, and iv) 256 variants with $\overline{\mathrm{H}} = 0.812$.

Different involutions $\mathbf{F}_{2/2}$ can be constructed using ten variants of the 2×2 substitutions, that are shown in Fig. 3 as elementary transformations $(x_1, x_2) \rightarrow (y_1, y_2)$. The following examples of the $\mathbf{F}_{2/2}$ elements correspond to the first subclass: (e,e,g,h); (e,f,j,h); (f,e,g,i); (e,e,j,i); (f,f,j,h); (e,f,j,i); (e,i,f,g); (e,g,f,h); (f,j,f,h); (f,i,e,g). For constructing the DDO boxes $\mathbf{F}_{32/96}$ and $\mathbf{F}'_{32/96}$ we have selected the (e,i,g,f) element that is described as follows:

$$y_1 = vzx_2 \oplus vx_2 \oplus vx_1 \oplus zx_1 \oplus z \oplus x_2; \qquad \mathrm{NL}(y_1) = 4;$$
$$y_2 = vzx_1 \oplus vz \oplus vx_2 \oplus zx_1 \oplus zx_2 \oplus x_1; \qquad \mathrm{NL}(y_2) = 4;$$
$$y_1 \oplus y_2 = vzx_1 \oplus vzx_2 \oplus vz \oplus vx_1 \oplus zx_2 \oplus z \oplus x_1 \oplus x_2; \quad \mathrm{NL}(y_1 \oplus y_2) = 4.$$

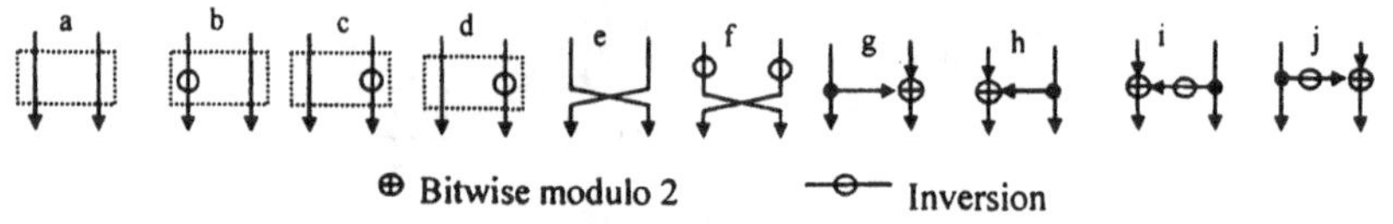

a b c d e f g h i j

⊕ Bitwise modulo 2 —⊖— Inversion

Figure 3: The 2x2 substitutions implementable by the the $\mathbf{F}_{2/2}$ elements

In Eagle64 cipher we use the DDO boxes $\mathbf{F}_{32/96}$ and $\mathbf{F}'_{32/96}$ presented in Fig. 4a and 4b as a cascade of four boxes $\mathbf{F}_{8/24}$ and $\mathbf{F}'_{8/24}$ (see Fig. 4c and 4d).

In the boxes $\mathbf{F}_{32/96}$ and $\mathbf{F}'_{32/96}$ the $V_1, Z_1, V_2, Z_2, V_3, Z_3$ components of the 96-bit controlling vector $V = (V_1, Z_1, V_2, Z_2, V_3, Z_3)$ are distributed in different ways. In the box $\mathbf{F}_{32/96}$ they are distributed from top to bottom, while in $\mathbf{F'}_{32/96}$ from bottom to top. Due to this fact, for the fixed value $V = (V_1, Z_1, V_2, Z_2, V_3, Z_3)$ the boxes $\mathbf{F}_{32/96}$ and $\mathbf{F'}_{32/96}$ perform mutually inverse transformations, i. e. $X = \mathbf{F}_{32/96}(\mathbf{F'}_{32/96}(X))$ and $X = \mathbf{F'}_{32/96}(\mathbf{F}_{32/96}(X))$ $\forall\ V \in \{0, 1\}^{96}$.

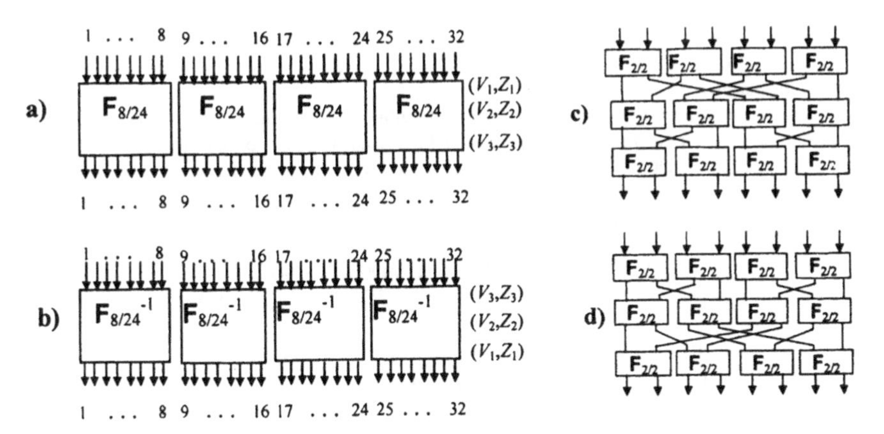

Figure 4: The DDO boxes $\mathbf{F}_{32/96}$ (a), $\mathbf{F'}_{32/96}$ (b), $\mathbf{F}_{8/24}$ (c), and $\mathbf{F'}_{8/24}$ (d)

3. Eagle64: A cipher for efficient FPGA implementation

The main feature of the Eagle64 cipher is combining the CSPNs with SPNs. The $\mathbf{F}_{32/96}$, $\mathbf{F'}_{32/96}$, and $\mathbf{F}^{*}_{16/16}$ boxes represent the part corresponding to the CSPNs. The $\mathbf{F}^{*}_{16/16}$ box comprises one layer of eight CEs $\mathbf{F}^{*}_{2/2}$ of the (e,b,b,c) type. The elements $\mathbf{F}^{*}_{2/2}$ are described by the following BFs:

$$y_1 = vzx_1 \oplus vzx_2 \oplus vx_1 \oplus vx_2 \oplus zx_1 \oplus zx_2 \oplus z \oplus v \oplus x_2;\quad \mathrm{NL}(y_1) = 2;$$

$$y_2 = vzx_1 \oplus vzx_2 \oplus vz \oplus vx_1 \oplus vx_2 \oplus zx_1 \oplus zx_2 \oplus x_1;\quad \mathrm{NL}(y_2) = 2;$$

$$y_3 = vz \oplus v \oplus z \oplus x_1 \oplus x_2;\quad \mathrm{NL}(y_3) = 4.$$

They have lower non-linearity than CEs (e,i,g,f), but they breed one ($\Pr(ijk) = \Pr(101) = 0.5$) or two ($\Pr(201) = 0.5$) active bits in the transformed data block while one active bit is fed into the controlling input. The Eagle64 comprises some elements of the Feistel cryptoscheme: i) the same algorithm is used to perform both the data encryption and the data decryption, ii) in the round transformation one of two enciphered data subblocks influences the transformation of the other one (due to the use of the DDO-box operations). General encryption scheme and round transformation (procedure **Crypt**) of Eagle64 are presented in Fig. 5.

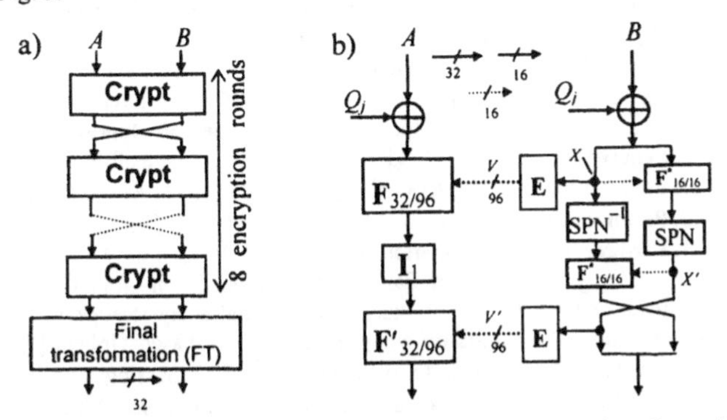

Figure 5: Iterative structure of Eagle64 (a) and design of procedure **Crypt** (b)

Two mutually inverse boxes SPN and SPN^{-1} shown in Fig. 6 are used in the right branch. The 4×4 substitution S (that is involution) specified in Table I is the main building block of the used SPN and SPN^{-1} boxes. To get good differential characteristics of the S-box it has been derived using the formula

$$G(z) = z^{-1} \bmod p(z),$$

where $p(z)$ is an irreducible polynomial. We have used the polynomial $p(z) = z^4 + z + 1$.

Subkeys $K_i \in \{0, 1\}^{32}$ of the 128-bit secret key $K = (K_1, K_2, K_3, K_4)$ are used directly in the procedure **Crypt**, as round keys Q_j (encryption) or Q'_j (decryption) specified in Table II. Thus, no preprocessing the secret key is used. More over, in each round transformation we use only one 32-bit subkey. The round key Q_j is simply combined with both the left and the right data subblocks. This makes the hardware implementation to be more efficient. Procedure **Crypt** is not involution. It is performed after combining the round key with data subblocks. In order to symmetries the full ciphering procedure a very simple final transformation (FT) is used, which is based on XORing a subkey with both data subblocks. Due to FT in Eagle64 the same algorithm performs both the encryption and the decryption, while different key scheduling is used for the two processes. The $\mathbf{I}_1$ permutational involution is described as follows:

(1)(2,9)(3,17)(4,25)(5)(6,13)(7,21)(8,29)(10)(11,18)(12,26)
(14)(15,22)(16,30)(19)(20,27)(23)(24,31)(28)(32)

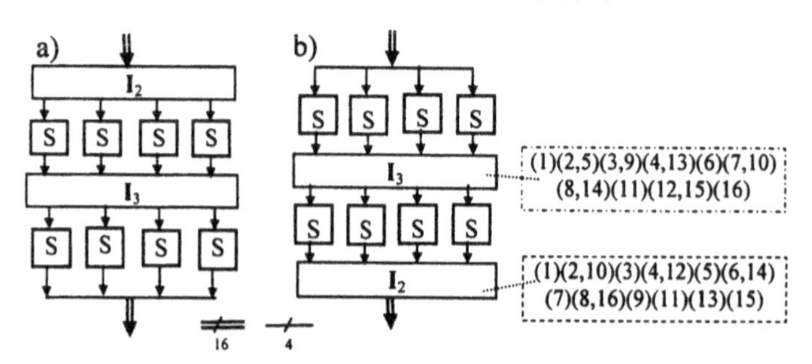

Figure 5: Design of mutually inverse operations SPN (a) and SPN^{-1} (b)

Table I: Specification of the S box

X	0	1	2	3	4	5	6	7	8	9	10	11	12	13	14	15
Y	0	1	9	14	13	11	7	6	15	2	12	5	10	4	3	8

Table II: The key scheduling in Eagle64 (j = 9 corresponds to final transformation)

Round number j =	1	2	3	4	5	6	7	8	9
Encryption Q_j =	K_1	K_2	K_3	K_4	K_1	K_2	K_3	K_4	K_2
Decryption Q'_j =	K_2	K_4	K_3	K_2	K_1	K_4	K_3	K_2	K_1

The 96-bit controlling vector of the $\mathbf{F}_{32/96}$ operation is formed with the extension box $\mathbf{E}$ described as follows: $\mathbf{E}(X) = V = (V_1, Z_1, V_2, Z_2, V_3, Z_3)$, $V_1 = X$; $Z_1 = X^{>>>2}$; $V_2 = X^{>>>6}$; $Z_2 = X^{>>>8}$; $V_3 = X^{>>>10}$, $Z_3 = X^{>>>12}$, where $X^{>>>b}$ denotes cyclic rotation of the word $X = (x_1, \ldots, x_{16})$ by b bits, i. e. $\forall i \in \{1, \ldots, 16 - b\}$ we have $y_i = x_{i+b}$ and $\forall i \in \{17 - b, \ldots, 16\}$ we have $y_i = x_{i+b-16}$. The 16-bit controlling vector (V_1, Z_1) of the $\mathbf{F}^*_{16/16}$ operation is described as follows: $V_1 = (x_1, \ldots, x_8)$ and $Z_1 = (x_9, \ldots, x_{16})$. The encryption process of Eagle64 could be described as follows:

1. For j = 1 to 7 do: $\{(L, R) \leftarrow \mathbf{Crypt}^{(e)}(L, R, Q_j);\ (L, R) \leftarrow (L, R)\}$
2. Perform transformation: $\{(L, R) \leftarrow \mathbf{Crypt}^{(e)}(L, R, Q_8)\}$
3. Perform final transformation: $\{(L, R) \leftarrow (L \oplus Q_9, R \oplus Q_9)\}$

Our preliminary security estimation of Eagle64 has shown that its four rounds are sufficient to thwart differential, linear and other types of attacks. Differential analysis appears to be more efficient than linear attack confirming the results on analysis of the DDP-based ciphers [5-7]. Our best DCs described in terms of [12] corresponds to difference (δ^L_2, δ^R_0) passing two (four) rounds with probability $\Pr(2) < 2^{-30}$ ($\Pr(4) < 2^{-60}$). For random transformation the probability of this characteristic is $\approx 2^{-55} > \Pr(4)$.

Investigation of statistic properties of Eagle64 has been carried out with standard tests, which have been used in [13] for testing five AES finalists. Obtained results have shown that three rounds of Eagle64 are sufficient to satisfy the test criteria, i. e. Eagle64 possess good statistical properties like that of AES finalists.

4. Hardware Implementation & Comparisons

We have implemented Eagle64 using FPGA Xilinx Vitrex Device and the loop unrolling architecture (denoted as LU-*N*, where *N* is number of the unrolled encryption rounds [14]; the iterative looping architecture corresponds to LU-1). This architecture has been selected to perform comparisons, since it suits well to implementation of the CBC (Cipher Block Chaining) encryption mode that is more interesting for practical applications. Due to the use of the FPGA-oriented primitives the Eagle64 is significantly more efficient for the FPGA implementation against majority of the block ciphers (for example, 3-DES [15]; SAFER+; IDEA [16]) including the DDP-based ones (Cobra-H64 [7], CIKS-1 and SPECTR-H64 [6]). Eagle64 implementation is also more efficient than the AES finalists (Rijndael, Serpent, RC6, and Twofish). Table III compares implementations efficiency (estimated as ratio "Performance/Cost" [14] and "Performance/(Cost·Frequency)") of Eagle64 with other well known ciphers.

Table III: FPGA LU-*N* Architecture Hardware Implementation Comparisons

BLOCK CIPHERS	Block Size	R	N	Area, #CLBs	F (MHz)	Rate (Mbps)	Integral efficacy Mbps/ #CLBs	Mbps/ (#CLB·GHz)
Eagle64 (Prop)	64	8	1	343	142	1,050	3.06	21.5
Cobra [7]	64	10	1	615	82	525	0.85	10.4
SPECTR [8]	64	12	1	713	83	443	0.62	7.5
CIKS-1 [6]	64	8	1	907	81	648	0.71	8.9
DES [13]	64	16	-	189	176	626	3.21	18.2
3-DES [13]	64	3×16	-	604	165	587	0.94	5.7
IDEA [14]	64	8	1	2,878	150	600	0.28	1.87
Rijndael [17]	128	10	1	3,552	54	493	0.138	2.56
Rijndael [12]	128	10	1	3,528	25.3	294	0.083	3.3
Rijndael [12]	128	10	2	5,302	14.1	300	0.057	4.4
Rijndael [13]	128	10	-	2,257	127	1,563	0.69	5.4
Serpent [12]	128	32	8	7,964	13.9	444	0.056	4.0
RC6 [12]	128	20	1	2,638	13.8	88.5	0.034	2.4
Twofish [12]	128	16	1	2,666	13	104	0.039	3.0

In addition two comparison models are used: Performance/Area and Performance/(Area·Frequency), since the first one is preferable while implementation of the ciphers using the same type of the FPGA devices, wherereas the second one is preferable in some cases while compared implementations use the FPGA devices of different types. It is obvious that the proposed applied methodology of Eagle64 achieves higher throughput values than other 64-bit ciphers and many 128-bit ones (except Rijndael implementation propose in [17]). It also needs significant less area resources than all other ciphers except the DES cipher [13].

5. Conclusions and Outlook

This work focuses on a new 64-bit block cipher design and implementation. We have advanced the DDO-based approach to design fast cipher. The present paper introduces new types of DDOs based on the use of the $\mathbf{F}_{2/2}$ CEs that are non-linear primitives. Besides, the $\mathbf{F}_{2/2}$ elements have significantly better DCs than switching element $\mathbf{P}_{2/1}$. Additionally, a new DDO-based cryptoscheme has been applied in Eagle64 providing extremely efficient FPGA hardware implementation. Eagle64 design combines the SPNs with CSPNs providing simultaneous transformation of the both data subblocks. It is characterized by significantly less depth of the round transformation than DDP-based ciphers [1,2,9]. Analogously to other DDP-based cryptosystems this cipher uses very simple key scheduling. The Eagle64 can be successfully used in both wired and wireless communications world, implemented on silicon with high speed performance and minimized allocated resources.

References

[1] A.A. Moldovyan and N.A.Moldovyan, A cipher based on data-dependent permutations, *Journal of Cryptology*, 15, 1, 61-72, (2002).

[2] N.D. Goots et al, *Modern cryptography: Protect Your Data with Fast Block Ciphers*, Wayne, A-LIST Publishing, (2003).

[3] Ch. Lee, D.Hong, Sun. Lee, San. Lee, S. Yang, and J. Lim, A chosen plaintext linear attack on block cipher CIKS-1, *Springer-Verlag, LCNS 2513*, 456-468, (2001).

[4] Y.Ko, D.Hong, S.Hong, S.Lee, and J.Lim, *Linear Cryptanalysis on SPECTR-H64 with Higher Order Differential Property*", *Springer-Verlag, LCNS 2776*, (2003).

[5] N.D. Goots et al., Fast Ciphers for Cheap Hardware: Differential Analysis of SPECTR-H64, *Springer-Verlag LCNS*, 2776, pp. 449-452, (2003).

[6] N. Sklavos et al, "*Encryption and Data Dependent Permutations: Implementation Cost and Performance Evaluation*", *Springer-Verlag, LCNS 2776*, pp. 337-342, (2003).

[7] N. Sklavos, N.A. Moldovyan, and O. Koufopavlou, High Speed Networking Security: Design and Implementation of Two New DDP-Based Ciphers, *Mobile Networks and Applications, Special Issue on: Algorithmic Solutions for Wireless, Mobile, Ad Hoc and Sensor Networks, MONET Journal, Kluwer*, 10, 219-231, 2005.

[8] N. Sklavos, and O. Koufopavlou, Architectures and FPGA Implementations of the SCO (-1,-2,-3) Ciphers Family, *International Conference on Very Large Scale Integration, (IFIP VLSI SOC '03)*, Germany, (2003).

[9] M.A. Eremeev, A.A. Moldovyan, and N.A. Moldovyan, *Data Encryption Transformations Based on New Primitive*, Avtomatika i Telemehanika, Russian Academy of Sciences, (2002).

[10] N.A. Moldovyan, M.A. Eremeev, N. Sklavos, and O. Koufopavlou, New Class of the FPGA efficient Cryptographic Primitives, *IEEE International Symposium on Circuits & Systems (ISCAS'04)*, Canada, (2004).

[11] A.A. Moldovyan, N.A. Moldovyan, and N. Sklavos, Minimum Size Primitives for Efficient VLSI Implementation of DDO-Based Ciphers, *IEEE Mediterranean Electrotechnical Conference (MELECON'04)*, Croatia, (2004).

[12] Moldovyan N.A. Fast DDP-Based Ciphers: Design and Differential Analysis of Cobra-H64, *Computer Science Journal of Moldova*, 11, 292-315 (2003).

[13] B. Preneel et al., *Comments by the NESSIE project on the AES finalists*, http://www.nist.gav/aes, (2002).

[14] A.J. Albirt, W. Yip, B. Ghetwynd, and C. Paar, FPGA Implementation and Performance Evaluation of the AES Block Cipher Candidate Algorithm Finalists, *3rd Advanced Encryption Standard Conference*, USA, (2000).

[15] B. Preneel *et al.*, *Performance of Optimized Implementations of the NESSIE Primitives*, Project IST-1999-12324, (2003).

[16] Cheung, O.Y.H, Tsoi, K.H., Leong, P.H.W., and Leong, M.P Tradeoffs in Parallel and Serial Implementations of the International Data Encryption Algorithm, *Springer-Verlag, LCNS 2162*, 333-347, (2001).

[17] C. Chitu, and M. Glesner, An FPGA Implementation of the AES-Rijndael in OCB/ECB modes of operation, *Microelectronics Journal, Elsevier Science*, 36, 139-146, (2005).

Brill Academic Publishers
P.O. Box 9000, 2300 PA Leiden,
The Netherlands

Lecture Series on Computer and Computational Sciences
Volume 2, 2005, pp. 191-196

A Reconfigurable Linear Feedback Shift Register Architecture for Cryptographic Applications

P. Kitsos[1] and N. Sklavos

Electrical and Computer Engineering Department
University of Patras, Greece

Abstract: An reconfigurable Linear Feedback Shift Register (LFSR) architecture for Galois field $GF(2^m)$, where $1 < m \leq M$, is presented. The value m, of the irreducible polynomial degree, can be changed and thus, can be configured and programmed. The value of M determines the maximum size that the LFSR can support. The advantages of this LFSR are (i) the high order of flexibility, which allows an easy configuration for different field sizes, and (ii) the low hardware complexity, which results in small area.

Keywords: Cryptography, elliptic curve cryptosystem, stream ciphers, bit-serial multipliers

1. Introduction

Multiplication over $GF(2^m)$ Galois field has many applications in cryptography [1]. For example Elliptic Curve Cryptography (ECC) requires GF multiplication. As the multiplication is very costly in terms of area and delay, a lot of research has been performed in designing small area and high-speed multipliers [2, 3, 4].

Previous GF multipliers could be classified into three categories: the bit-serial multiplier [2], the bit-parallel multiplier [3] and the hybrid multiplier [4]. Due to the infeasible complexity of the bit-parallel multiplier architecture in large Galois fields, hardware designers use traditional bit-serial multiplier. In addition, the usage of hybrid multiplier which use composite exponents in $GF(2^{nk(\cdot m)})$ is discouraged for ECC [5]. The traditional bit-serial multiplier uses Linear Feedback Shift Register (LFSR) [8] for modulo reduction purposes.

Finally, LFSRs are widely used in Pseudorandom Number Generators and Stream Ciphers [6].

In this paper, a small area reconfigurable LFSR architecture over $GF(2^m)$, where $1 < m \leq M$ is introduced. The degree of the irreducible polynomial m can be easily changed according to the application requirements. The maximum degree of the irreducible polynomial is M and the minimum is 2.

[1] Corresponding author, e-mail: pkitsos@ee.upatras.gr

The paper is organized as follows: in Section 2 a brief description of the fixed length LFSR is given. In Section 3, the proposed reconfigurable LFSR architecture is presented. The implementation measurements are shown in the sections 4. Section 5 concludes the paper.

2. Linear Feedback Shift Registers (LFSRs) Design

Fig. 1 shows a typical internal-XOR LFSR [7]. *n* denotes the length of the LFSR, i.e the number of Flip/Flops (F/Fs) in the register and $C_1, C_2, \ldots, C_n$ are the binary coefficients.

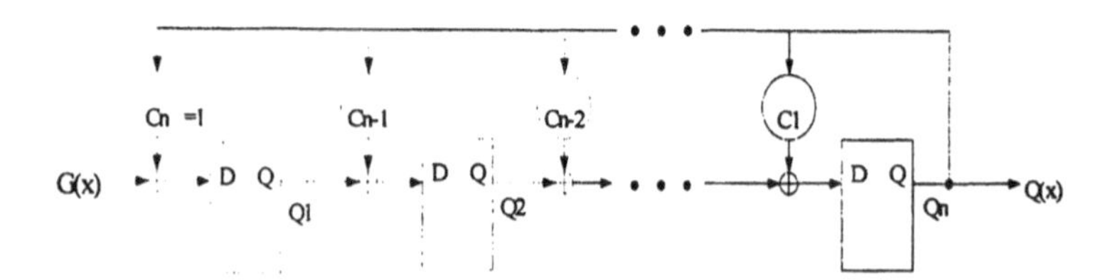

Fig. 1: A Typical Internal-XOR LFSR

The characteristic polynomial of this LFSR is,

$$P(x) = 1 + C_1x + C_2x^2 + \ldots + C_{n-1}x^{n-1} + C_nx^n \quad (1)$$

When C_i=1, implies that a connection exists. On the contrary, when C_i=0 implies that no connection exists and the corresponding *XOR* gate can be replaced by a direct connection from its input to its output. In addition, another polynomial of LFSR called reciprocal characteristic, is defined as,

$$P^*(x) = 1 + C_{n-1}x + C_{n-2}x^2 + \ldots + C_nx^n \quad (2)$$

The LFSR input sequence can be represented by the polynomial *G(x)* and the LFSR output sequence by the polynomial *Q(x)*. The first input bit enters to the LFSR corresponds to the coefficient with the highest degree of the polynomial *G(x)* and, the first output bit exits from the LFSR corresponds to the coefficient with the highest degree of the polynomial *Q(x)*.

In hardware terms, $G(x)$ (eq. 3) is the binary data stream to the modulo-2 division circuit, $Q(x)$ is the binary data stream produced at the output of the modulo-2 division circuit, and $R(x)$ is the remainder left behind the modulo-2 division circuit,

$$G(x) = Q(x)P^{*}(x) + R(x) \quad (3)$$

$P^{*}(x)$ is the reciprocal characteristic polynomial of the LFSR.

3. Proposed Reconfigurable LFSR Architecture

The proposed reconfigurable LFSR that can be used for variable field size m is shown in Fig. 2.

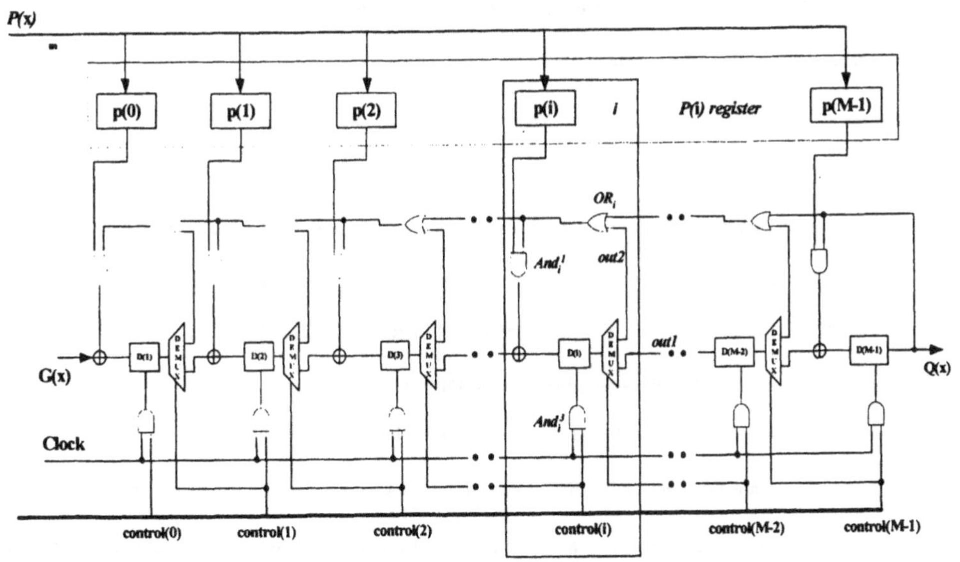

Fig. 2: The Proposed Reconfigurable LFSR Architecture

The proposed hardware implementation consists of a bit-sliced Linear Feedback Shift Register (LFSR) and is very similar to the conventional LFSR of Fig. 1. It requires M extra demultiplexers and M extra OR gates. Each slice i, consists of two subfield multipliers (*AND* gates), one subfield adder (*XOR* gate), one 2-output demultiplexer, one *OR* gate, and 2 one-bit registers ($P(i)$ and $D(i)$). The non-zero coefficients $p(i)$, of the irreducible polynomial $P(x)$, configure the LFSR, through the *OR* and *AND* gates of the feedback path. The maximum value of the field size is M, and it is determined by the application requirements. Each coefficient

$p(i)$, of the irreducible polynomial $P(x)$, is stored in the $P(i)$ position of the $P(i)$ register. If an irreducible polynomial of degree m with $m<M$ is required, the remaining $P(j)$ bits of the register are filled with zeros, where $m < j \leq M$.

Signal $control(i)$ selects one of the two demultiplexer outputs. The value of each signal $control(i)$ is defined as:

$$control(i) = \begin{Bmatrix} 1 & if\ i \leq m \\ 0 & if\ m < i \leq M \end{Bmatrix} \quad (4)$$

In the positions where $control(i) = 1$ ($out1$ is selected) the slice is active whilst if $control(i) = 0$ ($out2$ is selected) the slice is inactive. Inactive means that this slice is not used during the operation. *out2* forces the global feedback path with the proper value through the OR_i gate.

When the application requires operation with variable field sizes, the value of the degree m, and the set of coefficients of polynomial $P(x)$ are configured (i.e. programmed).

After m clock cycles the right output sequence $Q(x)$ is produced and the reminder $R(x)$ is stored in the register $D(i)$. The minimum clock cycle period is determined by the delays of the feedback path and the 2-output demultiplexer. It is equal to $2T_{AND}+T_{XOR}+T_{NOT}+(m+1)T_{OR}$, where T_{AND}, T_{XOR}, T_{NOT}, T_{OR} is the delay of the 2-input *AND*, 3-input *XOR*, inverter, and 2-input *OR* gates, respectively.

During computation each one-bit register $D(i)$, is controlled by the corresponding $control(i)$ signal. So, all the one-bit register $D(j)$, are set inactive by discontinue their clock signal. Thus, the unnecessary and wasteful transitions at all the one-bit register $D(j)$ are eliminated. This gated clock technique results in a significant power dissipation reduction.

It is suitable for polynomial multiplier architectures [8], for Pseudorandom Number Generators and new Stream Ciphers that uses LFSRs with variable field size.

4. Measurements

The area hardware resources and the execution time of the proposed LFSR hardware implementation are shown in Table 1. Since the proposed LFSR is the first reconfigurable architecture there are not comparisons with previous designs.

Table 1. Area hardware resources and execution time

# AND	m
# XOR	m
# REG	2m
# OR	m-1
# (2:1) DEMUX	m
Critical Path Delay	$2T_{AND}+T_{XOR}+T_{NOT}+(m+1)T_{OR}$
# CLK	m
Reconfigurable (Support any irreducible polynomial)	Yes

In applications with limited computational power, a hardware acceleration of the field arithmetic is necessary to reach high performance, especially if the irreducible polynomial degree *m* is beyond 200. The ECC with key size of 106 to 210 bits can yield the similar security level with the key size of 512 to 2048 bits of RSA algorithm. The key sizes are considered to be of equivalent strength based on MIPS years needed to recover one key [9].

In Table 2 measurements for the above mentioned binary field size 2^m are illustrated. The proposed LFSR is implemented in a Field Programmable Gate Array (FPGA). In this implementation the maximum field size *M* is 210. The polynomial degree *m*, varies from 106 to 210 bits.

Table 2. Proposed LFSR performance measurements

Binary Field Size (m)	FPGA Frequency (MHz)	LFSR Execution Time (usec)
106	33.5	3.1
119	30	3.9
132	26	5
158	22.4	7
193	18.5	10.4
210	17.1	12.3

5. Conclusion

A reconfigurable LFSR architecture is proposed in this paper. The LFSR is reconfigurable because it can operate with variable Galois field degree *m*. This LFSR can support any arbitrary irreducible polynomial. The result is computed after *m* clock cycles. The advantages of the proposed architecture are the high order of flexibility, which allows an easy configuration for variable field size 2^m, and the low hardware complexity, which results in small area. The proposed LFSR architecture is suitable for area-restricted devices and can be efficiently adopted from bit-serials multipliers, security systems and stream ciphers that use LFSRs with variable field size.

6. References

[1] A. J. Menezes, P. C. van Oorschot, and S. A. Vanstone, Handbook of Applied Cryptography, CRC Press, 1997.

[2] H. Li and C. N. Zhang, Efficient Cellular Automata Based Versatile Multiplier for GF(2^m), Journal of Information Science and Engineering, Vol. 18, No. 4, pp. 479-488, July 2002.

[3] Ç. K. Koç, B. Sunar, Low-Complexity Bit-Parallel Canonical and Normal Basis Multipliers for a Class of Finite Fields, IEEE Transactions on Computers, Vol. 47, No. 3, pp. 353-356, March 1998.

[4] C. Paar, P. Fleischmann, and P. S.-Rodriguez, Fast Arithmetic for Public-Key Algorithms in Galois Field with Composite Exponents, IEEE Transaction of Computers, Vol. 48, No. 10, pp. 1025-1034, October 1999.

[5] N. P. Smart, How Secure Are Elliptic Curves over Composite Extension Fields?, Proc. of the Advances in Cryptology - EUROCRYPT 2001, Innsbruck, Austria, May 6-10, 2001, pp. 30-39.

[6] B. Schneier, Applied Cryptography, Protocols, Algorithms, and Source Code in C, John Wiley & Sons 1994.

[7] M. Abramovici, M. A. Breuer, A. D. Friedman, Digital Systems Testing and Testable Design, IEEE Press, New York, 1990, pp 432- 448.

[8] P. Kitsos, G. Theodoridis, and O. Koufopavlou, An Efficient Reconfigurable Multiplier Architecture for Galois field GF(2^m), Elsevier Microelectronics Journal, Vol. 34, Issue 10,October 2003, pp. 975-980.

[9] A. K. Lenstra and E. R. Verheul, Selecting Cryptographic Key Sizes, Journal of Cryptology, Vol. 14, No. 2, pp. 255-293, 2001.